James F. Drane

A Liberal Catholic Bioethics

Ethik in der Praxis/Practical Ethics

Studien/Studies

herausgegeben von

Hans-Martin Sass

(Ruhr-Universität Bochum, Georgetown University Washington)

Schriftleitung: Arnd T. May

Band 15

LIT

James F. Drane

A Liberal Catholic Bioethics

LIT

Umschlagbild: Erwin J. Löhr

∞

Gedruckt auf alterungsbeständigem Werkdruckpapier entsprechend
ANSI Z3948 DIN ISO 9706

Bibliographic information published by the Deutsche Nationalbibliothek
The Deutsche Nationalbibliothek lists this publication in the Deutsche Nationalbibliografie; detailed bibliographic data are available in the Internet at http://dnb.d-nb.de.

ISBN 978-3-8258-6877-2

A catalogue record for this book is available from the British Library

Fresnostr. 2 D-48159 Münster
Tel. +49 (0) 251-620320 Fax +49 (0) 251-922 60 99
e-Mail: lit@lit-verlag.de http://www.lit-verlag.de

Distribution:
In Germany: LIT Verlag Fresnostr. 2, D-48159 Münster
Tel. +49 (0) 251-620 32 22, Fax +49 (0) 251-922 60 99, e-mail: vertrieb@lit-verlag.de

In Austria: Medienlogistik Pichler-ÖBZ, e-mail: mlo@medien-logistik.at

In Switzerland: B + M Buch- und Medienvertrieb, e-mail: order@buch-medien.ch

In the UK: Global Book Marketing, e-mail: mo@centralbooks.com

In North America by:

Transaction Publishers
New Brunswick (U.S.A.) and London (U.K.)

Transaction Publishers
Rutgers University
35 Berrue Circle
Piscataway, NJ 08854

Phone: +1 (732) 445 - 2280
Fax: + 1 (732) 445 - 3138
for orders (U. S. only):
toll free (888) 999 - 6778
e-mail: orders@transactionpub.com

ETHIK IN DER PRAXIS / PRACTICAL ETHICS
STUDIES - CONTROVERSIES - DOCUMENTATION

Ethik in der Praxis / Practical Ethics presents interdisciplinary *Studies*, discourses in *Controversies*, and *Documentation* for the analysis and evaluation of ethical principles and values and their integration into real-life scenarios of action and responsibility. Conceptual and methodological aspects of technology assessment and value assessment play an ever increasing role in public and academic debate, also in regulation, legislation, and political and business leadership, nationally and internationally.

In particular, research and application of communication technology, molecular genetics and biomedical technologies nationally and globally have become a prime topic for ethical, cultural, and transcultural assessment. Values have consequences, and technologies have consequences as well. Value assessment and technology assessment does not only belong into the ivory towers of academia or the lobbying floors of economic or political power. They belong to the people. They need to be present and vivid in public discourse and in individual and collective decision making. Values and virtues, not just technologies, have also been manipulated for purposes of exploitation, power and domination by political and moral authorities, by churches, ideologies, and various schools of thought.

Value assessment and technology assessment represent one of the driving forces in open societies of educated and responsible citizens, rich in a diverse spectrum of values, visions and virtues. Transdisciplinary research and analysis prepares and supports value assessment and technology assessment by individuals and communities, to provide material and clarification on controversial issues for public debate and public assessment of modern technology, its promises, its challenges, its temptations, and its cultural, ethical and technical risks.

The ethical and cultural assessment, acceptance, and risk of biomedical developments and of medical and clinical applications of new technology is a major focus of the *Ethik in der Praxis / Practical Ethics* series: the moral and immoral use of technologies at the beginnings and ends of human life, the use of contraceptives, antinidatives, hormone replacement therapies, or anti-aging and sexual performance drugs in loving partnerships, molecular genetic prediction and preventive medicine, drug safety and drug development, pharmacogenetics and gene therapy, health literacy, health promotion, and patient health care competence, health care and health responsibility, institutional ethics in health care facilities and nursing homes, justice in health care delivery, unfortunate or unjust distribution of health, of care and of wealth, advance directives and surrogate decision making, care for the environment and care for health, new individual and collective challenges in the ethical and cultural understanding of the beginnings and ends of morally to be recognized and legally to be protected human life, ethics in psychiatry and psychiatric research.

A Liberal Catholic Bioethics by James F. Drane reviews the entire field of modern biomedical ethics and its applications together with the new developments and options for diagnosis, intervention, and therapy in modern medicine from a position of Roman Catholic faith. But at the same time, it reviews critically the answers given by church authorities under the aspect of humanness and compassion and the Christian tradition of being a good neighbor in serving the frail, the needy, the sick, in enabling fellow humans to use technological and medical progress for living a more healthy, a more rewarding, a more civilized, ethical and cultured life.

Drane, a professor of Philosophy and a former priest, is not a Martin Luther and did not leave his church. He considers himself a faithful and devoted Catholic and understands Catholic reasoning as a service to the church and the Catholic tradition and future. The postscript positions him well into the actual dialogue within the church and with philosophical and religious moral reasoning outside the Catholic church, in the battle between hierarchical power and Christian thought and deed. The Introduction outlines his intention to further the dialogue, which much too often has been suppressed or ignored by those in power and hierarchy.

This book is a timely publication for all, who are searching for Christian and Catholic guidance in personal matters related to life and death, to health and disease, but also for a better understanding of the humane traditions of Christianity and medicine. It is a *study in controversies and a documentation of controversies*. May its careful and well informed analyses, reflections, questions, and arguments be discussed and not ignored, - on the way towards a more humane medicine.

Washington DC und Bochum, September 2010
Hans-Martin Sass

Content

DEDICATION

For Dr. James M. Gustafson, a friend and teacher who helped me to extend my theological views, deepen my respect for different perspectives, and form my ecumenical Catholic convictions.

ACKNOWLEDGEMENTS

Thank you Dr. Hans Martin Sass, a friend and colleague. This book is a work which you made possible. Dr. Pat Pineo took over the financial and administrative tasks of the James F. Drane Bioethics Institute. This freed me to work on the book and other writing projects. Thank you, Pat. Tina Paris provides secretarial help day after day and intelligent contributions to any work. Thank you, Tina. Finally, thank you Dr. Arnd May for your work on the text and for your support of this book project.

Preface

For persons of faith interested in ethics and bioethics there are particular problems created by the split within religious communities between conservative and liberal perspectives. What is the relationship between reason and authority in faith based moral teachings? Is there a distinctly Christian bioethics? Are all Catholics obligated to accept the official position of Catholic bishops on stem cell research which aligns the Church with the most conservative moralists? Can a truly humane medicine and a defensible bioethics be developed from a Catholic Natural Law perspective?

There are few influences today within the Catholic community that can unify thinking on pressing bioethics questions and heal the liberal/conservative split. The only reasonable thing to do, it seems, is to insist upon respect for different views and to use tough reasoning to advance honest dialogue. Maybe Catholics can take a few small steps toward healing the current splits in the Church by looking at some issues in contemporary medicine. In this book, I'll be trying to show that Natural Law, which provides the background theoretical vision in Catholic moral thought, can generate and justify a reasonable and respectable ethics, although different in some cases from official teachings.

I'll start by looking at the objective foundation of a humane medicine which is the doctor-patient relationship: the nature of illness and what doctors do to cure and to care for patients. Then I'll look at the moral issues of most interest to people; sexuality, birth control, abortion, physician assisted suicide and euthanasia, cloning, stem cell research, treatment for dying patients, relief of pain and suffering and finally how to pay for medicine today.

There is little doubt that the discipline of bioethics is the one which best reflects our era. It is the people's academic discipline. My hope is that students of bioethics will find some of the ideas which are developed here to be helpful. Approaching the end of an academic career, I would like to help people reach across dividing lines some common ground in moral issues that really matter.

DEFINITIONS

I. What is more Humane Medicine?

INTRODUCTION

Dissatisfactions with medicine are usually formulated in moral language. People talk about being unfairly treated, or about their rights being ignored, or about losing their dignity in the hospital. Patients have ideas about right and wrong in medicine, and often these do not coincide with what medical professionals understand as their duties and obligations. The ethical codes of doctors outline standards of ethical conduct as understood by fellow professionals, but such standards may not even mention the most pressing concerns of patients. The gap which separates patients from their doctors has become an ethical gap. Major ethical problems in contemporary medicine have to do with what traditionally has been referred to as the doctor/patient relationship.

Human illness is an enormously complex reality, and treatment is fragmented in today's medical system. Different specialists may be involved: surgeons and internists, oncologists and cardiologists, hospital staff physicians and private practitioners. Each specialist looks at the patient from his or her perspective. Often, doctors have trouble communicating with one another as well as with their patients. Certain basic patient needs frequently fall through the cracks. Patients can be left on their own trying to integrate information coming from different sources. Everywhere in today's high-tech medical system, communication about persons and among persons is difficult. The attention given to the doctor/patient relationship at best is superficial.

In many settings the doctor/patient relationship has been reduced to calling the patient Mr. and Mrs. instead of Bill and Betty. Lip service may be given to patient involvement in care, but in reality, information to the patient and patient consent may not be taken seriously. These important personal dimensions of medical care tend to be reduced to having the patient read a pamphlet and sign a form. Since no monetary compensation is provided for personal attention, it tends to be ignored.

Today's doctors are prone to evaluating themselves on the basis of objective technical competencies rather than by their personal care for patients. The surgeon who cuts straight becomes the model of physician excellence, and personal attention is left to lower-level health-care workers who know next to nothing about the patient's needs and have little or no authority. Many nurses on hospital floors are part-time workers who are hardly distinguishable from lower-level technicians. Nurses who are supposed to provide the humane personal care for high-tech specialists may not know the patients, or understand their situations. All nurses are under terrible pressure to get their basic paper work done.

Free-market capitalism has been inserted into medical care in order to reduce the cost. This change has caused further deterioration in the doctor/patient relationship. The care of patients has been turned into a commercial enterprise. What doctors historically committed themselves to do for patients is being lost. Professions are becoming jobs. Instead of a moral commitment to patients, doctors are guided by job descriptions written by administrators. The traditional ethics in medicine is being reduced to a set of technical job requirements. What historically had been a personal relationship with patients has become increasingly impersonal. Given this reality, it seems appropriate to look back in history and to review what have been for millennia the basic standards of humane medical care.

One solution being proposed to address the deterioration of modern medical care is an abandonment of the heart and soul of medical practice - the doctor/patient relationship. The proposed remedy for increased depersonalization of medical care is to abandon the personal and the humane in order to focus on the societal and the economic. Traditionally, a doctor is committed to doing what is best for a patient and from this commitment is generated an individual patient's trust in his or her doctor. This traditional model of the doctor-patient relationship and the corresponding trust which is generated is the background supposition which operates in professional codes, medical legislation and court decisions about medical mishaps. And yet, for some, it would be better to abandon this long tradition in favor of a social contract with patients which more adequately addresses the organizational aspects of health, especially its economic dimension (the need to control health care costs).[1] In the alternative model the way a patient is treated would depend upon a social contract which is faceless and impersonal. But illness always has a human face. It creates real and pressing needs in patients and calls for real and caring responses from medical professionals. Humane medicine always depends upon an individual doctor/individual patient relationship.

TRADITIONAL MEDICAL ETHICS

Historically, medical ethics has been expressed in codes. Statements about general duties and specific obligations of doctors were developed by learned members of medical brotherhoods who were responsible for regulating professional behavior. Because society recognized the right of professional persons to regulate their own behavior rather than have it regulated from outside, what doctors determined to be their rights and duties were, by and large, accepted by the larger community. As physicians changed their interpretations of themselves and their roles in society, they changed their ethical standards. Traditionally, medical ethics has been a self-declared and self-imposed ethic, outlining what noble service to others entails.

It would be an exaggeration, and I believe a mistake, to say that the traditional medical ethic is no longer valid.[2] The image of the good doctor incorporated in medical codes continues to have a positive influence on professional behavior, and many patients continue to accept and expect the self-imposed professional standards. But this is not as true now as it once was. More people today have a negative attitude toward the authority and privilege of doctors. A globalization of the democratic spirit has created an increased insistence upon participation by all people in rules or standards which affect them. Support for a self-regulating ethic, developed by trusted professionals to whom society cedes authority to determine their own responsibilities toward others, is weaker now than it has ever been. All this suggests the need to reconsider the foundations for what would be a more humane medicine.

THE NEW FOUNDATION

What constitutes the physician as a professional person is a special personal relationship with another human being who is ill. This bond between the medical person and the person who is threatened by the assault of illness on his or her very existence is both real and strong. In fact, it is the ground or foundation of medical ethics. Basic obligations of doctors (and nurses) are derived from the nature of this special relationship with sick persons. Only by appreciating what it means to be ill can doctors begin to understand what is ethically required of them as medical caregivers. Ethics in this sense is a lived set of obligations derived from a felt commitment to other persons. It is an ethics based on the relationship between doctors and patients and essentially an ethics of virtue.[3]

Illness can be understood in many ways. Modern medicine has understood this complex state with a model provided by science and with the language of its own diagnostic categories. This medical model is under critical fire from many sides, but it endures because it provides an academically sound and therapeutically helpful approach to illness. But no model of understanding exhausts the reality it attempts to grasp. By focusing on a scientific/physiological understanding, medical professionals do not exhaust the reality of illness as it is lived by the sick person. From the perspective of the person who is afflicted, the illness is an impairment which makes the whole person call out for help. Medical ethics forms its sense of duty from a personal commitment to respond to the call of another person who is in need. This is the starting point for a more humane medicine.

Medical ethics in this view derives its norms and obligations from the meaning of being ill and from a relationship between doctor and patient. It is the "sick experience" that gives rise to and is the very reason for being a medical professional. What the doctor should do (ought to do, is obliged to do) derives from the needs of the ill person. This view of medical ethics has its source in the professional's commitment to the sufferer. We can understand just what duties and obligations emerge from this commitment only when we appreciate the structure of illness. The key to a more humane medicine is not so much the duties and obligations which medical professionals decide to impose on themselves, but those which derive naturally from the concrete features of illness and the relationships between medical professionals and the needs of ill people.

Because society has an interest in its doctors and contributes considerable resources to their development, state licenses are required to practice medicine. With a license, rights and responsibilities are conferred. A relationship between the doctor and society has a lot to do with what traditionally was understood as medical ethics. Without denying this source of ethics, a more humane medicine takes a different reality as the source of professional obligations. The wellspring of medical obligation and responsibility is not so much the state or the professional organizations but the very humanity of the person who has been threatened and diminished by illness. Just how is the ill person threatened and diminished? Just what are the needs of ill persons? How can these needs best be met? Before being able to say how a more humane medicine plays out, these questions must be addressed.

THE NATURE OF ILLNESS

Serious illness is a broad insult to the whole human being. The accustomed equilibrium between the person and the world is rudely interrupted by symptoms, injuries, disabilities or disfigurements. The losses associated with these changes are many. Immediately the person is set apart and bears a burden. Characteristics which make human beings what they are the patient now has in diminished form. Illness wounds the core of human beings and diminishes life in its unique physical, spiritual, ethical and social functioning. Poverty and imprisonment also wound the core of human beings, but not as gravely as serious illness. A healthy person has power to deal with the deprivations of poverty and imprisonment, and this power is exactly what the sick person lacks. In serious illness, a broadside insult is suffered, and accompanying it is a devastating loss of power to remedy the insult.

LOSS OF PHYSICAL POWER

Because we are physical beings in a world which is not altogether friendly, human beings need power in order to survive. The world is often cruel because it responds neither to human needs nor human wants. Rather, it represents forces so powerful that all human skills and strengths are required to preserve life in such a world. When human beings fall ill, either physically or emotionally, the power they need to oppose hostile forces is undermined. Existence is threatening enough under ordinary conditions; but when serious illness strikes, the threat intensifies. Ordinarily the body is a willing instrument for resisting outside threats, providing satisfaction, and working toward desired goals, but in illness it becomes weakened and pained and therefore threatened. Illness deprives human beings of satisfactions as well as protection. Sick people can no longer pursue goals, play, enjoy and create. Frightened and diminished people look to the doctor for help much as a helpless child looks to a parent.

LOSS OF SPIRITUAL POWER

Human beings are more than physical beings. They have non-physical dimensions, like the capacity to grasp meaning, to pursue meanings, and to live in their own meaning systems. Victor Frankl[3] has shown that the loss of meanings is as much a threat to human beings as the loss of physical strength. The symbolic consciousness and symbol systems of human beings set them apart from other animals. Only for human beings do things in the world appear with a name and a meaning which invites further questioning and understanding. Basic meanings and interpretations give life coherence and a direction. They define a human being's experience, relationships and even sense of self.

Illness abruptly disrupts the ordinary meanings by which we human beings make sense of ourselves. All of a sudden a change in circumstance occurs, which forces a change in the sense of who we are. No longer do we live in the meanings created by our work and activities. Suddenly, our whole life becomes focused on this disruption. We find ourselves in an alien environment, dressed in strange clothes, confined to a bed which before we used only for nighttime sleep, and surrounded by people whose behavior and language are strange to us. Illness for a hospitalized or bed-ridden patient is the dominant reality. No wonder illness is so often associated with psychological upset.

It is difficult to give meaning to this new situation because most patients lack understanding of what is happening, why it is happening, what caused it, or how serious it is. They often do not know what treatments are available, the associated risks and benefits, the pain that lies ahead, whether or not they will survive. Besides disrupting a person's ordinary context of meaning, illness eludes a person's power to assign meaning, and consequently the patient suffers an upsetting disorientation caused by the loss of power.

LOSS OF ETHICAL POWER

Healthy persons have control over their behavior, are able to take initiatives and to make desired changes. Choice, control, initiative, imagining alternatives, decision making, responsibility, all these make up the package of human potential that is often referred to as human freedom. It is this freedom that condemns human beings to be ethical. Animals *adjust* to outside forces, but human beings must *make just* their responses to environment, by taking just initiatives. These human initiatives in turn forge the structure of a human soul.

If human beings can make themselves by their choices, it is just this characteristic capability that is threatened by illness. Sickness narrows the possibilities from which human beings make their crucial self-constituting choices. Not only are many life possibilities immediately cut down by illness, but patients cannot simply use will power to respond to the ill state. Pain and suffering weaken a person and force him or her to rely on the help of a powerful other in order to respond adequately and to regain some control. Patients cannot decide for themselves or act decisively in their own behalf as they did before illness. The freedom to make choices which are crucial to one's life is either lost or severely impaired by illness.

To make an effective choice about one's life, some vision of the future is required, and the loss of this vision is another of the casualties of illness. There is no reliable plan on which to base a decision, and the confidence required to carry through on a plan is also undermined. Illness is the enemy of agency and freedom and self-determination.

LOSS OF SOCIAL POWER

Human beings begin life in oneness, grow into separateness and then move back to new unity through tender, loving relationships. Besides being physical, spiritual and ethical beings, humans are also social in the sense of being constituted by a drive or a call to overcome separateness through relationship with others. A poster I saw recently in a doctor's waiting room showed a little girl hugging a kitten. The caption read, "It is a beautiful necessity of our nature to love and to be loved." One of the saddest and most frequently overlooked aspects of illness which requires hospitalization is rupture of the day-to-day contacts which are the heart of loving relationships. These familiar contacts are replaced by all-too-fleeting impersonal brushes with doctors, nurses, administrators, technicians and other members of the health-care team.

The most satisfying human relationships are those toward which a person is drawn by natural disposition and which are maintained by free decision. The relationships into which the patient is forced, however, are neither natural nor free. Illness forces a patient into relationships on the basis of helplessness. Doctors, nurses, and technicians have more power because of their skill. Besides, they speak their own technical language, which constitutes one more barrier to satisfying communication and relationship.

The loss of relationships is most serious in illness because the medical persons with whom the patient is forced into contact are often dominated by a reductionistic perspective on reality. By training, they look at illness in scientific terms and go about their work of applying the impersonal techniques of medical science: measurements, manipulations, analyses and interventions.

All is done with maximum objectivity. If medical professionals are trained to understand and treat disease scientifically, they are for that very reason vulnerable to forming only impersonal mechanical relationships. Friendship, concern, respect, trust - all these important values associated with humane caring are threatened in a modern doctor-patient relationship.

It is the depth of the loss associated with illness that makes the patient's relationship with the doctor so different from the relationship between lawyer and client, teacher and student, priest and parishioner. None of the needs or losses which give rise to other professional contacts approaches the loss one experiences in serious illness. Death is the loss of all that is characteristically human, and serious illness involves a threat of death. The whole person and not just an affected organ, tissue or cell is diminished in illness. In this threatened state, patients call out for help, and medical caregivers respond to this call. Doctors (and nurses) propose to help the sick person and not just to fix a malfunctioning part. If parts are repaired in however efficient a fashion, and in the process the person is badly treated, anger and disappointment result. The person who is ill calls for a human response and not just a technically efficient one. The reality of illness calls for a humane medicine.

THE OBLIGATIONS OF HUMANE MEDICINE

Once all the losses characteristic of illness are appreciated, one can see at least in outline form, the structure of a more humane medicine. The good doctor (or nurse) in this perspective is not just one who lives up to societal expectations, but one whose behavior addresses the losses suffered by the person with serious illness. The good doctor must strive to respond as one human being to another human being in distress.

Curing illness is important but even if curing is not possible, helping always is. The form which help takes is determined by all the ways in which illness deprives and diminishes the patient who calls for help. What specifically are the obligations of a more humane medicine which derives professional obligations from the needs of patients?

COMPETENCE AND COMPASSION

The doctor's first obligation is to acquire the requisite medical skill to provide the proper help. Competence means a capacity to diagnose, to treat, to cure and to ameliorate. Then there is the promise to put these skills to the service of a patient and to act in that patient's best interest. A profession is constituted by a public promise, and the substance of the physician's promise is to help persons who are ill. The current loss of public esteem and change in public attitudes toward medical professionals can be explained as a misunderstanding of the professional promise. Professionals have gradually moved toward a more restricted understanding of themselves which covers only technical competence. Patients, on the other hand, understand the professional commitment to extend beyond technique to help with all the losses associated with their illness.

Patients expect that the help promised by the medical professional will extend to what they understand as compassion. They expect the doctor to feel and to express feeling for them in their suffering. Scientifically oriented medical educators, however, insist on an ever-narrower technical competence that either leaves compassion to chance or assigns it to a social worker. But such a split cannot be right for medicine. Even the narrow medical specialist can show compassion. There is no reason why the understanding of medical competency cannot be shaped by the person who is ill rather than by some narrow view of organic disease. Humane medicine requires rootedness in a human reality.

The obligation to perform basic diagnostic and therapeutic functions competently is firmly established. Competence in the sense of technical competence will always be a basic obligation of the physician. No amount of compassion or sensitivity can make up for technical inadequacy. Medical professionals must be competent because they promise to help a patient and must be capable of keeping the promise. The patient must be given a reasonable chance of being restored to physical health, and the patient must be treated with compassion in the process.

DISCLOSURE AND PATIENT EDUCATION

The next important obligation corresponds to a patient's loss of meaning, which can only be addressed by a doctor. No one else can explain to the patient what is happening to him, why it is happening, how serious the illness is, and the different possibilities for treatment. A humane medicine based upon the needs of the patient, makes disclosure of information and patient education an important professional obligation. To make no effort to close the information gap between doctor and patient is to violate a basic obligation. Doctors who silently go about doing things to the patient without communication fail ethically, no matter how effective they are in what they do. Indeed, to ignore a patient's information needs is a form of cruelty. How much to inform, when to inform, what to do when information cannot be understood: all these are narrower technical questions grounded in the general obligation to help patients to understand when they are suffering from a loss of meaning.

PROTECTING PATIENT PARTICIPATION

If choice, initiative, decision making and responsibility are threatened by illness, medical professionals are duty bound to enhance patient decision making and self determination. Patients cannot be ignored when decisions are required which affect their very lives. Doctors may know what is best in a sense, but they cannot substitute their values or standards of decision-making for those of the patient. A humane medicine requires of the medical professional an effort to ascertain just how much initiative and self determination a patient wants to exercise and then to guarantee its expression. Patients cannot expect medical professionals to carry out their every wish, but they can expect from doctors help in making rational medical decisions. Illness undermines a person's voluntary capacities and sense of agency. Humane medicine obliges the medical professional to take these losses into consideration and to help patients to overcome them.

FRIENDSHIP

Everything about illness and its treatment in a modern hospital tends to separate the patient from close personal relations. Modern technical medicine treats patients as cases in a setting not unlike a modern factory. The efficiency of these treatment centers may be good or not so good, but in them the loss of personal relationships associated with illness is intensified. A humane medicine rooted in patient needs obliges the medical professional to a personal rather than a merely technical relationship with the patient. The one need most often overlooked is for a warm personal relationship with the healer. To this need corresponds what I think could appropriately be called the professional obligation of friendship. Patients need and want a person with whom they can share an awful experience, and it is just this sharing which constitutes friendship.

A relationship of some kind inevitably develops between a patient and the doctor, but most often it is not one of friendship. Some doctors today actually use patients as research subjects. On the other hand, a patient may merely use a doctor for reasons foreign to the goals of healing. While some merely functional relationships are inevitable and need not be ethically deficient, a functional relationship in marriage would be seriously defective; and the same is true in medicine. Not only does such a relational deficiency leave a basic patient need unmet, but it has a detrimental effect on medical treatment.

Treatment starts when the professional offers to assist the patient (*ad sistere* - to stay close to the other). Assistance in this etymological sense is an instance of medical friendship and is among the most powerful of all medications. Without it, powerful chemicals sometimes do not work. When it is present, placebos can become powerful agents. The patient needs the assistance and friendship of the doctor and nurse if other aspects of therapy can come to realize their best potential.

In an ethics based on patient needs, friendship becomes a major professional obligation. The required friendship is not like any other human bond. It has peculiar characteristics. Equality will never be one of its dimensions. It certainly will not last as long as most friendships. But it includes a real bond of affection. It is a form of love.

As with all friendships, a medical friendship is characterized by a proper balance between helping and letting be. The friendly professional does what is best for the patient but does not do so at the expense of the patient's freedom. The other must be respected in a friendship, and in medical friendships this translates into respect for the freedom of patients while at the same time helping them overcome their illness.

CONCLUSION

More humane medicine takes seriously not just illness but illness as experienced. When a person moves from a healthy state to one of serious illness, a profound change occurs which must be taken seriously. Illness has a structure of its own which can be understood, and this understanding can serve as the starting point for a Natural Law ethics of medicine which is more humane.

What we have seen of humane medicine are its basic ethical vision and an outline of the most basic professional obligations. What follows in this book is the application of a basic ethical perspective to specific areas of medicine and specific bioethical problems. Questions about practical ethical distinctions, concrete concepts and categories, specific behaviors and practices always have to be addressed in light of a basic background vision. The background vision or perspective will create a certain style of practicing medicine and a style of doing bioethics.

Style isn't everything, but it is important. Look at the man and the woman without style in order to understand this importance. Style gives tone and color to everything human beings do. The wrong style or absence of style can alienate persons and groups. Many complaints which people lodge today against medical professionals can be understood as complaints about style. People who deal with the public cannot ignore their styles of relationship, and this is as true of the medical professional as it is of priests or school administrators. To ground medical ethics in the structure of an illness experience and with a focus on the doctor/patient relationship creates a humane style of medicine and blends easily with a Catholic perspective in bioethics.

A more humane medicine roots medical ethics in the original reasons people decide to pursue a medical career. Doctors need to remember their motivation for entering the profession. Usually, they wanted to help people who were ill. Education in medicine and later medical practice in hospital settings imposes its own norms and its own patterns of behavior. The routines in medical residencies make it impossible for young doctors to spend much time considering the human condition of their patients. Once that exhausting period is over, however, it may be well to think again about original motivations. At some point in every profession, a return to roots is needed. Once in a while we have to look back to what we wanted to be, to our early ideals, and to our hopes and dreams for ourselves. If we do not remember where we started, or forget to check from time to time on the way we are proceeding, it is easy to lose our way.

POSTSCRIPT

What is true of doctors and medical professionals is just as true of Church officials and priestly professionals. Church officials, too, are influenced by a professional culture that easily can engender impersonal styles of behavior and cause priestly professionals to forget their early motives and reasons for choosing the priesthood. These earliest professions (priesthood and medicine) have much in common. A more humane medicine with only slight modification can serve as the outline for a more humane priestly ethics. A liberal Catholic perspective calls attention to these similarities.

Church officials focused on doctrinal orthodoxy and unchangeable moral teaching can be compared to the modern physician focused on organs and cells instead of the whole person. Both could be said to be concentrated on technological precision instead of on the persons whom they serve. In both instances, persons drop out. In one case it is the person of the patient. In the other it is the person of the ordinary believer.

Ignoring patients and ordinary believers violates the basic obligations of both the professional physician and the professional priest. Many patients today are alienated from doctors who have forgotten their ethical roots, and many ordinary Catholics are alienated from Church officials who have done the same. By one estimate, 20,000 Church communities of 1000 members each have been lost to the Church. Vatican officials put out continuous streams of disciplinary documents, doctrinal declarations and moral teachings. The official pronouncements from these priestly professionals insist on theological correctness; but like their fellow professionals in medicine, they tend to forget the needs of the people whom they are obligated to serve. Do they ever ask what effect the things which they say will have on ordinary believers? Do they care? Or have they forgotten the ethical source and basic principles of their priestly profession? Do they care only about theological correctness the way so many doctors care only about scientific correctness?

II. What is Liberal Catholicism?

DEFINITIONS

LIBERAL

The term "liberal" frequently is used to refer to liberal arts and to distinguish humanistic academic studies (philosophy, poetry, literature) from more technical and scientific ones (engineering, physics, chemistry). Broader definitions occupy a page length column in *Webster's Third New International Dictionary*.

Liberal, in religion, refers to thinking which is marked by openness and generosity. It refers to religious attitudes which are free from unnecessary restraints. A liberal religious person is neither strict nor rigorous, neither narrow-minded nor dominated by authoritarian viewpoints. "Liberal Catholic" to some may sound like an oxymoron. For many, it is a difficult concept to understand.[4]

CATHOLIC

To be Catholic means to be part of a community which goes back two millennia. It means to be a member of an institution which over the years, in interaction with all manner of cultural and political contexts, has developed distinctive practices, beliefs and moral teachings. For the liberal Catholic, these have to be respected, but respect involves grappling with the contexts in which the ritual practices, doctrinal beliefs and moral teaching were developed, in order to see how and when these might be differently understood in light of ongoing circumstances.

Non-liberal Catholics find change and recognition of the cultural contexts difficult to manage. Consequently, they get comfortable with the liturgical practices, doctrinal formulae and moral teachings of a particular historical period and commit themselves to maintaining them. Instead of trying to engage the inevitable process of change and development, they settle into the comfort provided by a myth of unchangeableness.

The liberal Catholic does not reject the doctrinal expressions or moral teachings or liturgical practices of the past. That would mean rejection of the Catholic community and its long history. But the liberal Catholic is interested in understanding changes which have occurred over that long past. This understanding protects against converting historical dogmatic formulae and particular moral teachings and liturgical or political practices into unchangeable and transcendent icons. Liberal Catholicism is more tolerant than non-liberal Catholicism. Taken to the extreme, conservativism becomes idolatry, and liberalism becomes lawlessness.

The moderately liberal Catholic tries to be an intelligent participant in a religious tradition. He or she tries to imitate the Jesus who was always willing to adjust laws and practices to meet the needs of vulnerable persons. The Jesus of the liberal Catholic is Jesus always being criticized and scorned by non-liberal types who believed that laws and rules which grew up in certain contexts could never be altered or changed no matter what the circumstance. The liberal's Jesus was always being accused and condemned by self-righteous types; and therefore, liberal Catholics try to avoid identifying their faith with condemning others or always being right. Jesus was not strict about all the laws and practices of His tradition. The liberal Catholic similarly tries to be more committed to loving and helping behaviors than to strict adherence to traditional practices and to always being right.

A liberal Catholic is often defined as one who rejects the authority of the Church in specific matters of doctrine and discipline, but who accepts the body of its teachings and its forms of worship. I'll accept that definition with certain nuances and reservations. It is not that liberal Catholics are cafeteria Catholics, picking and choosing the beliefs and practices which they like. Rather, liberal Catholics are committed to understanding history in order to understand and respond appropriately to the teachings and traditions of the Church. The Church lives in history and so do they. Sometimes this means being at odds with Church authorities.

Liberal Catholics like other liberals are committed to democratic political structures and the need to protect human rights. Current Catholic political structures are reflective of older, authoritarian, aristocratic, monarchical, political systems. Consequently, there is always a certain discomfort on the part of liberal Catholics with contemporary Church politics. Liberal Catholics want human rights recognized within the Church. They want a bill of rights for Catholics. They want rights respected, especially in academic life and with regard to women's issues. They want more open procedures when someone is accused or suspected of theological error. Liberal Catholics are embarrassed that the Church is one of the last institutions in the West still following pre-enlightenment politics and practices. Liberals want more convincing moral teachings and political structures.

LIBERAL CATHOLICS

a) John Courtney Murray

John Courtney Murray was a liberal Catholic (and my mentor when I was doing my doctoral dissertation).[5] He was a first class intellectual, a loyal Catholic and a dissident from the official teachings about Church relationship with politics: i.e., ideally, a state should endorse Catholicism and reject the false teachings of other religious groups. For centuries, the official moral teaching was that error has no rights. Murray rejected this idea and showed how historically it had led to scandalous practices. The official teaching, he showed, rather than advancing the teachings of Jesus, actually undermined His teachings and undermined the Church as well. The Church and state interact in the broader society but neither can dominate without undermining the freedom of citizens.

Murray argued at Vatican II for change and was successful. The Church abandoned the former teaching and endorsed religious freedom and respect for other beliefs. He showed that the Church was truer to its own beliefs in a religiously diverse society than in one where it was the privileged and established religion. Religion, separated from the state, influenced both state and society via pure reason. (Natural Law)

Father Murray reversed many of the ideas which Pius IX condemned in his infamous *Syllabus of Errors*. He reflected in his life and in his work the core of liberal Catholicism: that moral teachings have to be understood in their historical contexts and have to be open to development in light of better understandings of history and revelation. Even scripture must progressively be better understood by better understanding historical and cultural contexts.

Liberals, on the other hand, insist upon a Church with strong and continuing interaction with history and politics. Nothing in life, including religious life, is static. The Church's moral teaching and political structures are especially immersed in historical circumstances and have to be open to change in light of history.

b) John Noonan

John Noonan[6] is another example of a liberal Catholic. He took a stand on the issue of contraception. The traditional teaching was that Catholic couples can control procreations only by abstinence during fertile periods of the menstrual cycle. Of course, that did not work for most couples. A Catholic family consequently meant a big family. What Church officials neither understood, nor appreciated, was the pain and the disruption of family life caused by this teaching. Non-liberal theologians could only repeat the old formulas. In the meantime, Catholic couples suffered and families suffered, and many Catholics either left the Church or ignored the Church's moral teachings.[7]

Patty Crowley, a laywoman from Chicago who headed the Christian Family Movement with her husband, asked for input from Catholic women about birth control. She was overwhelmed by anguished letters about strained marriages because of forced sexual abstinence, ill health caused by multiple pregnancies, and disruptions of family structures under pressure from lack of birth control. This happened during the Vatican Council. She was one of the laypersons on a Papal Commission called by John XXIII, the liberal pope, to reconsider this teaching.

Another lay member of the Commission was John Noonan, who had written a scholarly book on contraception. His conclusion was that the issue was open to the possibility of change. For him, the key to reforming or updating this teaching was a better understanding of the philsophical, scientific and historical circumstances in which the traditional moral teaching was rooted. Non-liberal theologians on the Commission, however, could not appreciate either the experience of married persons or the scholarship of John Noonan. They could not imagine change because the older teaching was thought to be infallible and therefore unchangeable.

Noonan had argued that changing this moral teaching would strengthen rather than weaken Church credibility. It would make Catholic moral teachings about sexual morality more reasonable and more convincing. Instead of weakening the teaching authority of the Church, it would strengthen that authority. Treating this teaching as if it were immutable was for him and other liberals a blatant error.

John Noonan's thinking on contraception provides us with the paradigm of a liberal Catholic perspective. Liberal Catholics believe that the official Church record is full of complexities and ambiguities and cannot be turned into an unchangeable icon. The good Catholic, therefore, must work to move Church teachings and practices closer to Jesus' example by subjecting both to continuing critical, historical examination.

c) Garry Wills

Wills is another example of a liberal Catholic. Like John Courtney Murray and John Noonan, Wills is a faithful and practicing Catholic, and yet he has written an analysis of contemporary Church structures which makes clear the need for change. He shows in a recent book how an inflated, indeed sometimes idolatrous, image of the papal office has caused terrible damage to the Church.[8] There was always a special role in the Church structures for the Bishop of Rome. But over time we have seen an evolution toward a chair of Peter which sets one bishop over all others and to whom all others are required to pay divine-like homage. Besides destroying ecumenical hopes and needs, this inflated papacy has robbed the Church of honesty and consequently of credibility. An exaggerated importance of the role of the Pope first generated an assertion of infallibility. Then came a style of leadership and authority which neglected universally recognized human rights and operated with secrecy and dictatorial authority. This style generated a string of evasions, denials, irrational claims and an inability to tell the truth clearly.

For example, it took nearly 500 years to recognize that Galileo was right and the Church was wrong about the relationship between sun and earth. Even in this case, however, instead of an "admission of error," there was an explanation of what happened that never clearly admitted the mistake. The same thing happened with recent Papal documents on the Inquisition and the Holocaust. Instead of contributing to the formation of a community which reflects the honesty of Jesus, the modern imperial papacy has left us with a community in which liberal members are ashamed of the dishonesty of its leaders and take any authoritative proclamation with skepticism and suspicion.

Wills addresses the standard list of liberal Catholic complaints about the Church today: treatment of Jews, women, and Protestants; moral teachings about birth control; required celibacy for priests; totalitarian responses to critics; an authoritarian style of management; self protectiveness and secrecy. Wills looks to history in order to explain how these practices came to be and argues for reform. If things stay as they are and current practices are not modified, the Church, he argues, will suffer irreparable damage. He sees among other things a continuation of pedophelia-scandals followed by scandalous cover-ups. He predicts an increasingly homosexual priesthood.

In Will's liberal view, a core Church problem is the claim that Church authorities do not make mistakes (infallibility) and that things never change. Women, for example, in many ancient cultures were thought to be inferior, and consequently they were excluded from positions of authority and leadership in the Church. Now that such a belief is rationally unsustainable, new "reasons" are devised to offset pressures for change (e.g., only men were present at the last supper). Moral teachings and Church practices are not looked at historically and altered carefully in changed historical circumstances. Rather, they are kept in place by fabricating new reasons which are unconvincing. These shaky explanations are bolstered by an aggressive authoritarianism which does everything possible to keep change from occurring. Non-liberals refuse to look carefully and critically at the historical contexts out of which traditions arose. Instead they elevate traditions into unchangeable realities.

d) James Carroll

Another example of a liberal Catholic is James Carroll[9]. He is a novelist, a newspaper columnist, a National Book Award winner and a former priest. His book on the Church and the Jews has generated strong criticism and strong praise. What he did in *Constantine's Sword: The Church and The Jews: A History* is typical of a liberal Catholic. First, he addresses a shameful aspect of the Catholic Church: the perduring anti-Semitism which until just recently was reflected even in Catholic prayers and liturgies. Then he looks at history to understand how these sinful attitudes and practices came into being. Finally, he provides compelling reasons for real apology and real change.

Carroll's book about Jews in Christianity was triggered by the controversy about crosses erected at Auschwitz, and by the Church's attempt to apologize for the Holocaust in the Vatican document, "We Remember." John Paul II wanted to apologize. The document took about ten years to work through the Church bureaucracy, and when it finally appeared, it infuriated as much as it mended relations. Vatican officials could not admit the obvious truth that the Church was responsible in a substantial way for what happened to Jews under the Nazis. Consequently, they blamed what happened on sinful Catholics while insisting that the Church was blameless. Such blatant absurdity was a scandal to liberal Catholics and infuriated most Jews. So Carroll used this occasion to tell the truth about the Church's responsibility for two millennia of mistreatment of Jews.

Constantine's Sword argues convincingly that anti-Judaism is present, first in scripture (e.g., John's Gospel) and then in theology. He called attention especially to St. Augustine who probably has been the Church's most influential theologian (both Protestant and Catholic). Augustine condemned a slaughter of Jews which took place during his lifetime in Alexandria. "Do not slay them," he said. He advised rather that Jews remain within the Catholic culture as suffering witnesses to the evil of rejecting Church about Jesus.

Carroll's history explains the survival of Jews in Catholic cultures (many dissidents were eliminated) as well as their enduring suffering. Carroll's exposé of the place of anti-Judaism in Catholic Church history fills a liberal Catholic first with remorse and then with a commitment to remedy this evil rooted in Catholic history. It fills non-liberal Catholics with fury and disdain for a writer who would claim to be Catholic and yet say such things. Truth, even when it hurts, is liberation for the liberal and unacceptable to "true believers".

History for James Carroll both explains to Catholics who we really are and provides the foundation from which to construct effective Church reform. His historical explanation of anti-Judaism can have a reforming influence on the Church, both Catholic and Protestant. Non-liberals in both Catholic and Protestant communities have for centuries settled comfortably into their traditional prejudices and the convictions that they are right. Divided Christian communities may not make much ecumenical progress in working out who is right, but Protestants and Catholics could easily and immediately join in recognizing where we are both wrong. Since Christians of different convictions share the same failures in regard to treatment of Jews, they are united in moral failures and sinful ways. Maybe they could start moving toward unity through shared acts of contrition and penitential liturgies.

Being "right" in thinking or speaking typifies most non-liberal believers. Liberals who stand over against this mind-set, however, sometimes fall into the same absurdity. It is one thing to subject Church formulae and teachings and policies to historical examination and then to look at the need for change in light of new circumstances. It is another thing entirely to claim that every historical re-examination renders a total truth or that every recommended change will actually be for the best. Liberals can be just as authoritarian as those with whom they differ. Sound historical analysis is the foundation of liberal arguments for change, but history never comes over in clear unarguable insights. What we see in history is to some extent at least influenced by the instruments of interpretation which we bring to the research. Consequently, the liberal agenda for change has to be advanced with both humility and patience.

Carroll sometimes shows something less than humility in the repeated references to himself in his book, which is a history. And he doesn't show much patience. Typical of a Catholic liberal, he finishes his history with a call for reform (a call for Vatican III). In this regard, Carroll shows how vulnerable we liberals are when we set out to set things straight. He anticipates solving every major problem by adopting a full liberal set of perspectives and convictions. Some non-liberals don't want anything to change, but many liberals want everything to change in one grand liberal reformation.

James Carroll's solution to anti-Judaism in the Church is to revise what we believe about Jesus. He wants to move away from belief in Jesus as Messiah and away from the centrality of Jesus' death on the cross. But Christians can be freed from anti-Judaism without abandoning their convictions about the significance of Jesus' life and death. Jesus' life and death is not the source of anti-Judaism. We can cure Christian anti-Judaism without abandoning basic Church beliefs. Carroll's searching of his own soul for the sources of anti-Judaism can help every reader to carry out an effective cleansing. Then we need to keep belief statements open to refinements and changed emphasis. James Carroll, in typical liberal fashion, shows the importance of history for understanding both the Church and ourselves. At the same time he shows how history inevitably provides only limited insight and how conclusions drawn for historical constructs may be exaggerated and even wrong.

Revolutions are neither realistic nor helpful for Church reform. As a matter of fact, things are already changing and have been doing so since the Vatican Council and the document *Nostra Aetate*. Carroll's book will feed into a stream of other writings directed toward eradication of a 2000-year-old anti-Judaism blight.

There is an inevitable relationship of religion and politics. The joining of Christianity with the Roman Empire in Constantine's reign gave the Church both theological formulae and political structures. These included anti-Judaism, and this blight cannot be removed overnight. But historical revelations can be followed by personal self-discoveries and then by ecclesiastical reforms. That's the liberal Catholic agenda. And it can be carried out only by tough-minded critical history done by Catholics who care about the Church.

e) Sister Joan Chittister

One final example of a liberal Catholic shows the same tough-minded, critical engagement joined to a strong caring commitment to the Church. I am referring to Joan Daugherty Chittister, OSB.[10] Sister Joan holds a Ph.D. in philosophy and has been a distinguished teacher all her adult life. She is a rational Catholic in the Aristotelian sense: one who questions, looks for causes and tests all claims. She is a strong advocate of woman's rights in the Church, democratic structures, tolerance, ecumenism, anti-violence and anti-war. Her questioning has led to tension with Church officials. The latest incident involved the auxiliary bishop of Pittsburgh and the secretary of education for the diocese. These two Church officials forbade Catholic schoolteachers permission to attend an educational convention where Sr. Joan was scheduled to speak. The bishop tried to enlist other ultra-conservative bishops, (e.g., the bishop of Peoria, IL), to join him in his totalitarian tactic to force the convention organizers to cancel Sister's talk. The convention organizers refused to withdraw her invitation to speak. The convention had its largest audience ever. The bishop may have advanced his career goals in the Church hierarchy, but he helped people to see as well the difference between mean-spirited ultra-conservativism and intelligent liberalism. All liberal Catholics are not saintly and all bishops are not mean-spirited. But some are, and when the two perspectives are placed alongside one another, the liberal Catholic vision ends up being both edifying and convincing. The Bishop's behavior shows us the need for reform in the Church's authority structures and political practices. Sister's behavior shows us an imitation of Jesus' way.

Liberal Catholics question things, and questioning defines Sr. Joan. For her, the unquestioned life is unimaginable, as well as not worth living. Her life is immersed in the Church, and that life never ceases to wonder and to question. She handled the cowardly behavior of the bishop with grace because she has understood from long experience that her questioning threatens certain people and some institutions. Those who cannot handle questioning respond with hostility. She responded to the hostility with grace because she knows that courage is required to face questioning and that not everyone is courageous. Given the present Church structures and style of authority and the historical treatment of women, she knows that questioning is dangerous.

Reflecting her philosophical formation and her respect for Aristotle, the greatest of all philosophers, she understands her humanness in terms of a questioning engagement with reality. Life without questioning is not worth living because it fails to engage the unique feature of human being. Questioning, looking for causes, testing hypotheses, these distinguish us as humans and protect us from leading wasted lives immersed in untruth. Without truth there is no integrity to the individual self, and whole societies can become blind instruments of evil if questioning is repressed. The questioning which permeates Sr. Joan's life is not aggressive and is not disrespectful. She questions Church teachings and practices that cry out for better understanding and for change. Without a place for questioning in the Church, we are all condemned to darkness. Sr. Joan's questioning shines light. Those who would repress her are advocates of a Catholic Church of blind robots.

To question Church teachings and practices for a person whose life is immersed in the Church means to question the very foundations of one's own life. She is a professed religious; and when she questions the beliefs and practices of the Church, she questions everything about herself. But only to a person who questions herself and her Church will certain flaws and failures be revealed.

Sr. Joan started questioning at the tender age of six when her widowed mother brought her new stepfather to the priest for instruction. The priest told Mr. Chittister that only Catholics can go to heaven. Coming from a good Protestant family, he responded with disbelief and asked, “Do you mean that my saintly mother didn’t go to heaven when she died?” The priest’s response was to shove into his face the statement “To be saved, it is necessary to be Catholic.” “You can read,” he said. “Read it yourself.” Her father’s reaction was one of anger. He thought that this insensitive priest was also crazy and he discontinued instructions. Whether the priest was mad or just stupid and discourteous, he lived and taught that incredible religious distortion because Catholics didn’t question. A little girl, her siblings and her mother’s life were changed because of a teaching that was both stupid and harmful. Did God send to hell the billions of human beings who preceded the coming of Christ? Does he do the same for most of the world’s human beings who are Protestant or belong to other faiths? How ridiculous! How scandalous a claim! And yet before questioning by faithful Catholics like Joan Chittister, this type of stupidity and the scandal it created prevailed.

This liberal Catholic woman started questioning as a little girl; and the questioning permeated her school life, religious life, family life, political life and community life. Questioning constructed her inner self. Questioning moved her from one point to the next in her religious life. She is not going to give up questioning, because questions continue to be raised by the whole reality which she confronts. The worst imaginable reaction to Sr. Joan and other Church liberals is the use of crude power tactics to silence them. She will not stop questioning any more than other faithful liberal Catholics will. Ultra-conservatives cannot eliminate the liberals. They have to learn to live with them respectfully within the same Catholic community. The same is true for Catholic liberals who have to try to understand and live in peace with ultra-conservative Catholics.

Sr. Joan is the paradigm of the good Catholic liberal. She is honest and pursues the truth both in her own life and in the life of her Church. She refers to herself as an Irish Roman Catholic woman religious. She has faced radical questions raised by each descriptive adjective. The questions she asked over the years changed her as a person. The questions she continues to ask of the Church are just as radical and just as likely to change the Church. Without liberals like her raising questions, the Church will still change, but the changing will have no direction and will not likely be for the best. Sincere and tough questioning anticipates real understanding and leads to positive change.

One guarantor of intelligent improvement in the Church is the cultivation of liberal Catholicism.

LIBERAL CATHOLOCISM AND HUMANE ETHICS

In looking at ethical issues in contemporary medicine and trying to suggest ways to make medicine more humane, I will be questioning many of the Church's official teachings. I will sometimes offer arguments which may be at variance which common understandings of Church policies or teachings. I will be arguing both theoretical and practical medical issues from a Catholic viewpoint.

Some of the positions defended here are actually traditional Catholic moral teachings, but ones now under stress because of pressures created by conservative Pro-life Catholics and Church bureaucrats (see "Mechanical Nutrition and Hydration for Advanced Alzheimer's Patients"). In other cases, I try to argue for a moderation of Church policies and suggest ways of supporting moderation on medical issues like abortion (e.g., "Modifying the Abortion Policy: The Role of Metaphor in Medicine and Law"). On broader issues of sexuality, the arguments support the convictions of most Catholics but still are opposed by Church authorities. Because so much of contemporary medicine is linked with powerful new technologies applied in medical care, I try to address the issue of limits to technological interventions in genetics and reproductive technologies. Then, since all ethical decision-making assumes background ethical theory, I try to argue for a Catholic understanding of Natural Law. I certainly do not claim to be right about all these complicated issues. What I want to do is to try to construct a defensible, Catholic *perspective* on medical issues for consideration by both believers and unbelievers.

My modest hope is that I'll be able to touch on certain critical themes which might help to make the treatment of vulnerable people a little more humane, and help alienated Catholics to see that there is a defensible perspective which supports their moral convictions based on solid experience.[11]

III. What is Bioethics? A History

DEFINITIONS

THE BEGINNINGS

No other field of study reflects the contemporary age more faithfully than bioethics, a systematic study of moral conduct in life sciences and medicine. Medicine and the life sciences are to our period in history what religion and salvation were in medieval times. They are the focus of enormous societal resources and the central concerns of most modern people. Bioethics pulls together under a single discipline the ethical dilemmas associated with contemporary bioscientific research and its application in medicine. What for centuries was referred to as medical ethics is now a part of the broader discipline of bioethics. This new discipline began only recently in the developed countries which had to face many new ethical challenges generated by the biosciences. But the same ethical problems now challenge people everywhere.

It's hard to pinpoint the beginning of an historical period or a cultural development or even an academic discipline. In most cases beginnings are too far in the past and become lost. Even when relatively few years have passed from the start of something new, initiating events may be diverse. Distinguishing a first step from preliminary or background influences is always problematic. To talk about the beginning of bioethics is inevitably to speculate.

Did bioethics as a new discipline begin with the founding of the Hastings Center and The Kennedy Institute in 1969-1970? Or was it the formation of an Ethics Committee in Washington State during the sixties, trying to set up ethical standards for the distribution of a scarce medical technology to dying patients (Renal Dialysis). Or perhaps a book by Van Rensselaer Potter[12] in 1971, called *Bioethics* launched the discipline. Or was it Protestant German Pastor Fritz Jahr who in 1927 coined the term and defined in contrast to Immanuel Kant's Categorical Imperative a *Bioethical Imperative*: 'Respect every living being as an end in itself and treat it, if possible, as such' Beginnings are rarely discrete, specific and easily identifiable realities.[13]

Advances in life sciences that gave impetus to the field of bioethics in developed countries now are part of contemporary life in developing nations as well. Modern high-tech medical centers can be found in major cities all over the world. People everywhere face the same ethical problems associated with human experimentation. The press in Europe and Latin American and Japan now gives the same prominence to ethical problems in medicine that we have seen for decades in the United States. Physicians in other countries are aware of the need to understand the ethical issues generated by modern medical practices and to update their professional codes. Foreign and domestic politicians anticipate direct government involvement in health-care regulation and this means involvement with ethical values, especially the principle of justice. In just a few decades bioethics has become a major discipline worldwide. Most likely, it will reflect the ethos of 21st century civilizations for future historians.

Because of its important place in contemporary society, bioethics has undergone meteoric changes in the last three decades. First bioethics centers, institutes, commissions and boards were established in the U.S. and Canada. European nations and the European community followed quickly with their own initiatives. Scholars from Japan and Southeast Asian countries who spent time in Canada or the U.S. or Europe returned to direct the establishment of bioethics institutes in their own countries. A bioethics program has been established in Santiago, Chile, which promotes development of the discipline throughout Latin America and the Caribbean. Bioethics conferences have already been held in Eastern Europe, and work has started on the development of bioethics centers there. Even recently independent countries in the former Soviet Union and Yugoslavia are organizing conferences on bioethical problems and planning bioethics institutes. International bioethics exchanges have begun to take place, and already the field is changing as a result of efforts to agree internationally on ethical rules and policies. An originally dominant North American-style bioethics is now changing under the influence of European, Asian and Latin American perspectives. There is an identifiable Catholic bioethics, as well as Protestant, Jewish, and Muslim bioethics.

In its initial stages bioethics was concerned with ethical issues generated by developments in medicine. Later the subject matter was broadened to include all the biosciences, but biomedical ethics remains a core part of this now larger field. Difficult as it is to identify precisely its actual beginning, several influences can be recognized as important contributors to the rapid development of the now paradigmatic discipline.

MEDICAL EXPERIMENTATION

German medicine in the nineteenth and twentieth centuries served as a paradigm for modern medical practice. It was linked to laboratory science which meant that mainline medicine required proven effectiveness based on rigorous experimentation which inevitably involved human subjects. The misuse of human subjects in medical experimentation created the first modern ethical crises and the first calls for a new medical ethics. The Nuremberg Code responded with what came to be one of the foundations of the new ethics, an informed consent requirement. Whenever medical professionals use human subjects for their research, they have to guarantee respect for each research participant.

The voluntary consent of the human subject is absolutely essential. This means that the person involved should have legal capacity to give consent;

should be so situated as to be able to exercise free power of choice, without the intervention of any element of force, fraud, deceit, duress, overreaching, or other ulterior form of constraint or coercion; and should have sufficient knowledge and comprehension of the elements of the subject matter involved as to enable him to make an understanding and enlightened decision. This latter element requires that before the acceptance of an affirmative decision by the experimental subject there should be made known to him the nature, duration, and purpose of the experiment; the method and means by which it is to be conducted; all inconveniences and hazards reasonably to be expected; and the effects upon his health or person which may possibly come from his participation in the experiment.[14]

The violation of traditional medical ethical standards by misusing patients created widespread moral outrage. Vulnerable, weak and needy human beings, instead of being cared for and protected, were used and misused. This called for a new set of ethical standards. Quickly the new ethical standards were extended from medical experimentation to medical treatment because vulnerable patients required protection there as well. Benefit balancing against risks and disclosure of dangers became as much a part of treatment ethics as of research ethics.

News of grossly unethical behavior by some Nazi physicians during World War II was followed in the United States by a series of revelations of similar ethical failures involving experimentation on vulnerable patients in American medicine (Willow Brook School, Jewish Hospital in New York, and the Tuskegee Syphilis Study). In 1966 Henry K. Beecher, a Harvard physician, published an article in the *New England Journal of Medicine* in which he exposed patterns of unethical conduct in medical research.[15] The misuse of human subjects by U.S. physicians and Beecher's comments on the misuse were widely publicized and contributed substantially to a growing public interest in a revised ethics of medicine.[16] Ethical failures associated with research had a major influence on this new field of study. Concern about ethics and experimentation is as strong today as it was at the beginning of modern medicine, and bioethical regulation of research now takes place around the world.

The imperative to make scientific progress in medicine is present anywhere contemporary medicine is practiced. Because the authority of physicians tends to be stronger in foreign countries than it is in the United States, conditions exist for similar failures everywhere. Only a well-developed and widespread bioethics could keep ethical tragedies generated by medical research from occurring. No society can afford to leave the balancing of individual patient rights with scientific progress solely a matter for medical scientists to decide. Standards for the conduct of human experimentation had to be imposed by outside authority everywhere modern medicine was practiced. This is true in the great medical centers certainly, but today, even community hospitals and doctors offices have become places where testings of medicines and other medical research occurs.

MEDICAL TECHNOLOGY

Flowing from government investment in medical science were all sorts of new medical technologies and therapeutic interventions. The linkage of medicine with science, which had begun in the late nineteenth century, began to pay off handsomely. New medicines, dialysis machines, organ transplant techniques, mechanical organ transplant support systems, medically delivered alimentation technologies, ICU's, lifesaving surgeries, etc. were developed.

Each new development created new ethical problems. In the 1960s, an Ethics Committee was formed in Washington State which tried to make ethically defensible decisions about who would receive dialysis when this scarce technology could not be provided to all. Bioethics was not just concerned with medical experimentation using human subjects. It was concerned with medical treatment and the participation of patients and communities in medical decision-making as well as with who would have access to the treatments.

Historically the medical profession always accepted moral responsibility for the exercise of physician power over patients. The medical profession expressed this responsibility in medical codes and association treatises. All socially authorized professional power requires public accountability, and this is especially true of medical professional power. The right to practice medicine is associated with moral restrictions on that practice imposed either from inside the profession or by the government. As medical practice became more powerful with technologies, ethical problems associated with medical practice proliferated.[17] The range of things physicians could do for patients expanded along with the effectiveness and intrusiveness of their technological interventions. Micro problems arose with each intervention. Macro problems generated by the relationship of technology and human life also had to be addressed. In both developed and developing nations, physicians became concerned with creating and updating their ethical codes.

Scientific and technological medicine moved medical treatment procedures into the public forum. Medical treatment began to take place principally in public hospital settings where ethical responses had to be publicly defensible. Twenty-first century technologies may make earlier therapies look primitive and uncomplicated, but we can see in early technological breakthroughs the driving force behind the new focus on ethics and the emergence of modern bioethics. The importance of health as a value and medicine as a discipline made biomedical ethics an important field of study almost everywhere.

Before the 1950s, "doctors know best" captured the attitude most people had toward medicine and summarized a traditional paternalistic ethics. After the Nuremberg trials and the increased influence of experimentation on practice, this older paternalistic ethics gradually gave way to different standards of right or wrong. Other attitudes, different norms, more and different principles coalesced to create a new medical ethics and the new discipline of bioethics. Once bioethics was born, it quickly developed.

GOVERNMENT INVOLVEMENT

After World War II, developed nations put great emphasis and large amounts of money into the medical field. Consequently greater attention had to be given to the ethical issues which inevitably accompany medical advances. In the United States, the U.S. Public Health Service, an agency of Health, Education and Welfare and later Health and Human Services, was made responsible for protecting the rights and welfare of human research subjects.

In the 1960s it promulgated ethical standards for the conduct of research. In the 1970s a National Commission for the Protection of Human Subjects of Biomedical and Behavioral Research was formed. It worked for four years and made 125 recommendations for improving protection for the rights and welfare of human subjects. The Commission published the *Belmont Report*, which identified the basic ethical principles (respect for persons, beneficence, justice) and provided philosophical justification for humane treatment of human subjects.

Subsequently, the federal government under the auspices of the same Public Health Service continued to update regulations and to require assurances of ethical compliance with them from any group carrying out research on human subjects. Government, through its support of medical projects, became not just a source of bioethics policy and rules, but it established commissions which articulated ethical justifications for a new approach in the conduct of medical practice. Government and its interests played a major role in the development of modern bioethics.

After the National Commission's work, the U.S. government continued its involvement in bioethics in the form of a new President's Commission for the Study of Ethical Problems in Biomedical Research. This commission was formed in the 1980s and mandated to provide reports to the President, the Congress and relevant departments of government in order to guide politicians in developing legislation. In addition, its work provided guidance to health professionals, health educators and the general public. The President's Commission published eleven volumes, nine reports, the proceedings of a workshop on whistle blowing in research, and a guidebook for local committees that review research with human beings. The work of this one government commission has had an enormous influence on bioethics. Some of the Council's works provide some indication of just how bioethics expanded in the first few decades: *Compensating for Research Injuries, Deciding to Forego Life Sustaining Treatment, Defining Death, Implementing Human Research Regulations, Making Health Care Decisions, Protecting Human Subjects, Screening and Counseling for Genetic Conditions, Securing Access to Health*

Care, Splicing Life, Whistle Blowing in Biomedical Research, Being Human[18], *Human Dignity*[19].

Government has continued to play an important role in bioethics. There are bioethics programs associated with all the major government research efforts in medicine. Part of the funding committed to medical research is usually allocated for bioethics. Sometimes government funding for controversial research is withheld as a way of exercising bioethical regulation. Federal bioethics legislation and regulations are ongoing.

NON GOVERNMENTAL INFLUENCES

Millions of government dollars contributed to the birth and expansion of bioethics. Modern bioethics, however, came from more than government initiatives. Non governmental institutes and centers also sprang up to respond to the pressing new problems generated by bioscience.

As early as the 1950s, the Institute of Religion at Texas Medical Center in Houston started working on ethical issues in medicine, and a Society for Health and Human Values was formed by religious thinkers interested in promoting the humanities in medical education. In the 1960s the first Department of Medical Humanities was started at Pennsylvania State University Medical Center in Hershey, PA, with a faculty weighted toward medical ethics. The next decade witnessed the appearance of the Hastings Center in Hastings, NY (late 1969, early 1970), and the Kennedy Institute of Ethics at Georgetown University (1971). Both these initiatives attempted to bring depth and rigor to the new discipline.

The Kennedy Institute model was university-based. It developed a National Reference Center for Bioethics Literature which in effect became the best library resource in the world for an expanding new literature. Its scholars came from many different disciplines, worked somewhat independently of one another and served as faculty for a Ph.D. program in bioethics at the university. One of the first scholars, Warren Reich[20], a Catholic theologian, put together the *Encyclopedia of Bioethics*, which became a major literary resource for the discipline. A Protestant scholar, Leroy Walters[21], started an annual *Bibliography of Bioethics* and developed "BioethicsLine," an on-line computer database. Tom Beauchamp[22] and James Childress[23] published the important book, *Principles of Bioethics*. As new areas of the expanding field of bioethics emerged, scholars from the new areas who were interested in ethics came to the Kennedy Institute to study, to write and to teach.

On the West Coast, bioethics was advanced by the writings and teachings of Albert Jonsen[24], at the University of San Francisco and the University of California in San Francisco. In Southern California, William Winslade, a lawyer, taught and wrote about the issues. In Chicago, Dr. Mark Siegler[25], a professor of medicine at the University of Chicago, started a training program for clinical bioethicists. In Texas, Dr. Tristram Engelhardt[26] promoted humanities in medical education and clinical bioethics. He worked at Baylor University Medical School with a Jewish scholar, Baruch Brody. On the East coast, Charles Culver, Bernard Gert and K. Danner Clouser cooperated to produce some of the first substantial books and articles about theory and methodology.

The Hastings Center was started by Daniel Callahan, a Catholic layman with a background both in philosophy and theology. At the Hastings Center scholars were brought together to work both independently and in groups in order to develop sound ethical policies for specific problems. The Hastings Center continues to publish policy recommendations and topical reports and to influence government responses both directly and indirectly. *The Hastings Center Report*, founded in 1971, carried articles on ethical issues in medicine, the life sciences, and the professions. Over the years, it became the most important journal in the new field.[27]

Since the early 1970s and the work of bioethics pioneers, literally hundreds of bioethics centers, programs, journals, and newsletters have sprung up. Every year the books and articles on bioethical subjects number into the tens of thousands. From a small and recent beginning, bioethics has become a major field of study. The American Hospital Association in 1987 published a description of 77 bioethics organizations. Since that time, the number of such organizations has tripled.

Government commissions, academic centers and non-governmental institutes combined to contribute to the development of bioethics in the United States. Increased interest on the part of professionals fed interested personnel into the growing number of bioethics education institutes. Academically based bioethics centers trained professionals for teaching posts in the new field. Hospitals sometimes hired their own bioethicists for education and consultation, thereby creating job opportunities for trained bioethicists. Bioethics committees were organized in health-care settings, and committee members needed education in a field now with an extensive literature. Attitudes of skepticism and resistance towards the humanities components in scientific medicine gradually gave way to acceptance on the part of faculties, students and professionals. The thousands of bioethics articles and books annually testify to what this field has become over the last few decades.

MEDICAL CULTURE

In the early days, bioethics was all about medicine but the discipline did not develop within medicine. Early in the 1970s it muscled its way into the medical culture, largely as a result of the efforts of one man, Dr. Edmund Pellegrino. Dr. Pellegrino was both a physician and a humanist, a practicing doctor and a bioethicist. His efforts show how government and non-governmental groups cooperated to launch a new discipline and an important new social enterprise. But for the combined effort of different forces, the field of bioethics would have been a social and a political influence, but would have remained marginal to actual medical practice. The story of how bioethics entered medical culture links Dr. Pellegrino with the Institute on Human Values in Medicine (non-governmental institute), the National Endowment for the Humanities (a federal government program) and with medical school faculty all over the country. It was Pellegrino who made bioethics part of the medical school curriculum.

More and more doctors became specialists and used the new technologies which developed after World War II. For the first time in medical history, doctors became strangers to their patients. And the moral assumptions which doctors and patients had shared for centuries (e.g., paternalism, beneficence, non-maleficence, confidentiality, and respect) were increasingly compromised. An increased number of moral questions arose from the use of the powerful new technologies. Journalists recognized these as issues of interest, and conflicts over the uses of these technologies were given prominent coverage in the press. What was the proper use of the new technologies: respirators, artificial nutrition and hydration, dialysis, artificial insemination, birth control devices? Could patients simply trust the doctor to do what was in their best interest, or did they have to insist on more personal freedom and a more adult relationship with the medical stranger who was using new technologies on them?

The new moral questions about doctors and patients and technologies were being asked at a time when traditional ethical values rooted in religion were being either revised or rejected. Questions posed by medicine, medical practice, dying patients, defective newborns, etc. had been addressed for centuries in Catholic moral theology because the Catholic Church was engaged in health-care ministries. But in the late 1960s and early 1970s, the moralists and ethicists who stepped in to think about all the new questions came from very diverse backgrounds. There were some Catholic theologians, but many of the early bioethicists were secular lawyers, Jewish doctors, Protestant philosophers and social scientists.

New secular ethical theories were proposed to justify recommended ethical directions. In addition to theoretical literature, the new discipline was engaged in the development of practical procedures for conflict resolution. Bioethics in effect was digging into the doctor/patient relationship, the core of the medical profession. Once this turn was taken, there was no way that bioethics could remain outside the strongly walled culture of medicine. Medical students had to be taught the discipline. Medical faculty members had to be trained. The medical curriculum had to be expanded to include this new discipline. The physician who saw this inevitability was the physician bioethicist, Edmund Pellegrino.

Dr. Pellegrino worked with members of the Institute on Human Values in Medicine and the Society for Health and Human Values. Many of the members were Protestant clergymen and chaplains who were in dialogue with physician members of their organizations. These persons recognized the changes which were taking place in the doctor/patient relationship and saw the need to train medical professionals in the new legal and ethical standards for this relationship. The operating background assumptions of this group of professionals were theological. Gradually the members merged with more secular groups like the Society for Bioethical Consultation and the American Association of Bioethics.

In 1969, the National Endowment for the Humanities, a federal government initiative, provided grant money for Dr. Pellegrino. The project was to train medical school faculty in bioethics and then to carry out visits to U.S. medical schools. These visits put the new discipline of bioethics on the ground in the middle of the medical world. The grant money also supported fellowships for the training of medical school students and faculty in bioethics. Physicians were enabled to study ethics, and ethicists were provided opportunities to become familiar with the hands-on culture of medicine. This effort by Dr. Pellegrino and colleagues resulted in major innovations in medical education. It started what later would become the separate branch of clinical bioethics.

This project brought bioethics teams to most U.S. medical schools (77 different locations). The team members brought organizational ideas, lectures, teaching techniques and encouragement to faculty and students to pursue the new discipline. They were able to influence both faculty members and medical school administrators. They held meetings with interested faculty and students. Gradually, the most influenced faculty became advocates for the new discipline within the medical school environment. As bioethics programs got up and running, return visits by team members provided evaluations of the schools' efforts and recommendations for improvement. Through this one effort, bioethics was able to move from academic reflection to practical changes in the way doctors handled their patients. Bioethical norms and values were given concrete expression in medical practice so that they became more than just social/legal/political standards.

THE LAW

When issues raised by experimentation and treatment could not be resolved at the patient-physician-family level, they were taken to the courts. The first court cases involved tragic situations with dying patients, e.g., the Karen Ann Quinlan case. Families and hospital staffs disagreed about whether to withdraw life-sustaining technologies, and courts were asked to make life-and-death decisions. The disputes attracted the media and created front-page stories. People wanted to hear about the tragic cases because they touched concerns and worries in every family. Court decisions in the highly publicized cases contained ethical arguments that themselves stimulated further ethical arguments. Later court decisions either approved or overturned earlier ones, and a whole corpus of legal bioethics literature came to be.

The combination of media attention and public interest made bioethics important to politicians who saw the need for creating statutes to defend patient and family rights in health-care settings. Every state now has laws covering bioethical concerns. New laws and new cases continue the interrelation between bioethics and law. Other nations are facing the same pressures, and lawmakers as well as judges everywhere look for help from experts in this new field in order to develop sound legislation.

In Europe and North America, the law adopted many of its positions from ethics. For centuries Catholic moral theology held that patients have a right to refuse any treatments, even life-sustaining treatments, if these were experienced as burdensome, risky, or costly. Statutory law and court cases upheld this tradition. And the influence went both ways. North American bioethics adopted standards for surrogate decision-making which were developed in the law: e.g., *subjective standard* (what the patient actually chose), *substitute judgment* (what the patient would have chosen) and then *best interest* (what is considered medically best for the patient). And the core bioethical requirement of informed consent came directly from case law.

Questions about proper treatment of patients or the proper form of a doctor/patient relationship were taken to court and what gradually accumulated were judicial decisions which set out new legal requirements for medical practice. Despite a platitude to the contrary, morality is legislated in the United States and elsewhere in the world. Judges and state and federal legislators established the foundation of a new medical ethics discipline by setting out, as early as the 1940s-1950s, ethical standards for medical practice. Law was definitely involved in the beginning of contemporary bioethics.

Ordinarily, successful law codifies custom, and this was true to some extent of the new bioethics legislation. Surgeons had traditionally sought patient consent for their dangerous interventions. The information they provided was aimed at helping patients to endure the agonizing pain of the surgery. What traditionally took place in the exchange between surgeon and patient, however, would not meet later standards of informed disclosure and voluntary consent. Almost any communication between the surgeon and patient satisfied traditionally understood consent, as long as what the doctor said was not untruthful and the patient gave some form of assent. Without any communication from the surgeon or consent from the patient, courts considered surgery to be unauthorized and a form of assault[28].

One court case in the middle 1950s actually used the term informed consent[29] to describe a legally imposed ethics for doctors. Later decisions built on the *Salgo* decision and gradually spelled out in greater detail the parameters of a physician's duty to make adequate disclosure to a patient. In 1960, *Natanson vs. Kline*[30] spelled out a standard for reasonable disclosure. The court required that the nature and possible consequences of a treatment be disclosed. Reasonable disclosure after *Natanson* meant that the required communication between doctor and patient address questions about risks, consequences and options; i.e., what reasonable persons want to know.

In the Natanson decisions, the legal system established a new ethical standard for the doctor/patient relationship which in turn became a bioethics cornerstone. Initially, this discipline focused on trying to justify the new legally created practices by using the principle of autonomy, and then by trying to settle questions like which risks, how much information, how many alternatives. However one defines the beginning of bioethics, court cases in the 1940s, 1950s and 1960s laid the groundwork for its initial focus and direction. American courts, however, were not the only background influence. International courts, as far back as 1948 (Nuremberg) had spelled out legal and ethical standards for informed consent in research which gradually were accepted as well for treatment procedures.

A SHIFT TOWARD SOCIO-ECONOMIC ISSUES

When the discipline which emerged out of U.S. culture spread to other parts of the world, the strong cultural influences at work in American bioethics become more evident. Once persons in Europe and Asia and Latin America became involved in bioethics, they reacted sometimes strongly to the U.S. emphasis on individual rights and on the principle of autonomy. The new discipline, it turned out, was imprinted with a U.S. cultural stamp. One of the fruits of a dialogue with bioethicists from other cultures was a move toward a broader bioethics which is still in its formative stages. Now, more than ever, bioethicists are aware of social, cultural, and religious influences on ethical reasoning.

During the early years, bioethics was dominated by clinical problems at the beginning and end of life. Those problems are still with us, but now it is not the withdrawal of life supports that is being argued (for example, the Karen Ann Quinlan case) but rather the active involvement of a physician in directly taking a patient's life (the Jack Kevorkian campaign). Genetic technologies and their associated dangers now dominate the field. Increasing attention especially is also given to professional medical ethical standards which can easily be compromised by health-care delivery systems which are market dominated. Socioeconomic factors today threaten to turn doctors and nurses into obedient employees rather than independent professionals.

I first started reading *American Medical News* years ago to keep up with the culture of the medical profession. Then, the paper was full of stories and articles about clinical issues. Now this weekly AMA newspaper is full of articles about economics: salary concerns of doctors, pressure on doctors coming from HMO administrators, problems associated with the management of a practice, Medicare and Medicaid policies, and how these affect physician working conditions, moves to approve collective bargaining units for physicians, the increasing dominance of free market capitalism in hospitals and nursing homes. These are the problems which now seem to preoccupy the medical profession. Correspondingly bioethics and bioethicists are increasingly involved with the socio-economic reality.

In the early years the principles of beneficence and autonomy were dominant, and bioethics was about balancing the two or choosing one over the other in a clinical dilemma. Now the discipline is more focused on justice and equality and on the structures of the health care delivery system. The dominant issue now is how to bring about fair access to health care. In order to do bioethics seriously when justice and equality and economic problems dominate, the bioethicist must know in depth the social sciences.

In an earlier book,[31] I took a strong stand in favor of limiting the issues addressed by an institutional bioethics committee to clinical care problems. Admittedly allocation problems or the socio-economic issues are every day more important. But the older ethics committees were not formed and trained to handle the allocation issues so dominant today. The same is true of most practicing bioethicists. Clinical bioethicists and health-care ethics committee members have to be familiar with clinical contexts, with contemporary medicine, with state statutes, court cases and government regulations which address clinical matters, as well as with the concepts and theories of bioethics which were designed to handle clinical subtleties. It is unrealistic, I believe, to expect these same persons to master a very different set of ethical concepts and theories, a totally different type of fact or data, and a completely different group of social sciences. To understand the socio-economic or justice issues in sufficient depth and to be able to offer realistic ethical guidance to institutions requires a differently trained bioethicist and a different type of ethics committee.

Human beings cannot be expected to be experts in every field. One of the worst dangers bioethics and its practitioners are exposed to is superficiality, joined to an inflated sense of authority. Once I attended an international meeting where a very inexperienced member of the bioethics community spoke about ethics committee responsibility. She had the committee solving life and death dilemmas in the intensive care unit, in addition to problems of the kitchen staff and maintenance personnel. The talk was plainly ridiculous and pointed up a danger: inexperience and superficial understanding of one's limits can create "ethics experts" ready to solve any and every conceivable problem as long as it is articulated in terms like good and bad, or right or wrong. In order to avoid this pitfall, bioethicists and ethics committee members should be either trained in clinical medicine and ethics or trained in health-care organization and social sciences, but not ordinarily both. Rare is the person or group of persons conversant with the complexities of clinical medicine and in addition, the background sciences, relevant data, applicable court decisions and all the different ways in which justice and equality can be applied in particular institutional or organizational crises.

The core issues for a socio-economic committee are related to what treatments a just health-care system should offer its patients, or what limits on treatment may ethically be imposed. No health-care system, whether private or public, capitalistic or socialistic, can offer every therapeutic option. Human beings can never be satisfied either in their demands for enough health or in their demands for enough defense. Military budgets and health-care budgets always have to set limits. Any company, indeed any government, would quickly go broke trying to satisfy every request or demand. Limits have to be set, and at some point someone has to say no. But what would constitute a basic fair minimum which everyone would have access to, and what could justifiably be denied? Medical help in a humane health-care facility is an obvious good which persons want, but it is not the only good.

Other goods, like education and police protection and a judicial system and roads and welfare, compete with health care for limited socio-economic resources. But how and where can we set the limits without violating a rightful demand for justice and equality? These are critical types of questions today but they are very different ethical questions from those earlier ones having to do with the removal of technologies at the end of life. The latter remain with us and have become more complicated. New clinical technologies and treatment possibilities raise all sorts of complex new clinical problems (e.g., genetic diagnosis and therapy). Some gifted bioethicists have managed to develop the background understanding to address the new and the old issues with expertise, but they are rare. I worry about less talented people who presume to handle all issues without adequate background training.

Bioethicists who presume to help a health-care system or a particular health-care organization to decide how much to offer, and what limits to impose, are faced with a monumental task. Systems and institutions have to be economically sound or everyone loses. But how do we make decisions about whom to cover or when to treat? Who decides to deny treatment? Do we do what is most just by providing treatments which have statistically the best outcomes or do we provide treatments for those who are the most disadvantaged? How do we determine best outcomes or greatest need? How do we avoid self-interests when we are being paid big money to provide the ethical advice?

Bioethicists who work for companies and capitalistic health-care institutions have their hands full. They are forced into areas which are addressed only with help from economics, accounting and political science. They are forced into areas which once were addressed in the philosophy of medicine: What is medicine? What are the limits of medicine? What are the objectives and goals of medical practice? Doctors cannot do everything their techniques and powers make possible. They cannot ethically amputate an arm to help a man get more Social Security benefits. But other questions about limits are more difficult and throw open tough economic issues and highly philosophical questions about medicine's reason for being. Can the first clinical bioethicists presume to know enough about social science, economics and philosophy of medicine to offer defensible responses to socio-economic questions today? Even to devise fair procedures for deciding such questions requires a specialized expertise. Strength of character becomes an issue when the bioethicist is being paid big money to aid the company or the HMO to remain economically profitable and avoid being sued.

And yet, if no qualified ethical reflection on these questions is available, then decisions will be made by administrators of health-care organizations, or politicians, or stockholders. Without serious bioethical reflection on socio-economic questions, pure greed and self-interests will permeate and dominate health-care decision making. Patients do not have to be professionally trained economists or accountants to notice when their interests are not being taken seriously. It is already obvious to most people that large amounts of health-care monies are being spent on advertisement, salaries for CEO's and payouts to stockholders. Everyone who has any contact with today's health-care systems notices new obstacles to care created by long lines and delays in access to appointments with specialists. Issues of fair access and just limits are critical, and yet they demand an ethics expertise which only persons who know economics and accounting and policies and the inner workings of the medical care delivery system can provide. They require a new type bioethicist with a different type academic background and enough economic independence or inner moral strength to avoid being prostituted. Making ethically acceptable socio-economic public policy for companies or hospitals is not the same as making discrete clinical decisions about individual patients at the end of life.

Both type questions, however, fall under today's broader definition of bioethics. They are ethical questions raised by the practice of medicine. They require virtue and character as well as analytic and procedural ethical abilities. They emerge in health-care systems and institutions and involve both patients and health-care professionals. And both types of bioethics questions can be generated by particular cases. What to do with this and that patient can be the initiating event for both types of ethical reflection. In socio-economic cases the questions would be whether the patient or the condition can be covered. Responding to cases in a consistent and objective manner gradually creates a direction for an institutional ethics policy. Over time and with added expertise, ethics policies can be modified. Setting the limits of medical care is too important an issue to be excluded from the new discipline or from today's bioethics committees.

Helping medical practitioners with clinical problems characterized the first steps of the discipline and helped to define bioethics. Since its beginning, other problems have emerged. The discipline has expanded. And yet, ethical problems associated with treatment or nontreatment of particular patients in a clinical setting cannot be set aside. Continued effort has to be expanded to refine both the norms and the procedures for deciding about clinical cases. Bioethics can become as abstract and theoretical as any other academic discipline, but it started out as an applied ethics, and it cannot abandon this continually important activity without losing its soul.

THE FUTURE OF BIOETHICS

The academic efforts of bioethicists to address the dilemmas characteristic of modern medicine provided crucial assistance to societal leaders, both political and professional. But society and societal leaders were not the only ones to benefit. Ethics itself was benefited by bioethics. In 1973 Stephen Toulmin wrote an article about how medical ethics had saved ethics from decline and disinterest.[32] The problems with which medical ethics grappled not only created a new interest in ethics but also saved ethics from an irrelevance created by an overly abstract, rationalistic, linguistic approach. Philosophers, theologians, lawyers and social scientists suddenly found the ethical aspects of medicine and biosciences to be areas of fascination and started studying and writing about them. Today newspapers carry daily and weekly columns on ethical questions generated in contemporary life.

Will the astounding expansion and central importance of bioethics continue into the twenty-first century? A quick and clear response to this question comes from considering two recent issues: the genome project and AIDS. The human genome is mapped and as the information locked in human genes is opened, the ethical problems generated by this new information have exploded. Data banks of individual DNAs are being established. Government agencies, police, employers and insurance companies, if they gain access to the data, could substantially influence human lives. The information developed by this one biomedical project has ominous as well as hopeful potential. Only with well-thought-out ethical standards and judiciously developed ethical policies can the worst results imaginable be avoided. The very dignity and freedom of human life swings in the balance between ethical and unethical handling of this one bioscientific project.

The genome project was the life science project of the 1990s and can easily be compared with the physics project to unlock the atom in the 1940s. The potential for good is great, but unless the associated ethical issues are openly discussed and thought through in advance, human life as we know it today in a civilized, free and democratic society may be undermined. The sheer numbers of ethical complexities are hard to imagine, but the ominous consequences of not attending to them are even more ominous. With good reason some portion of the money allocated for the genome project is committed to bioethics. Ethical questions generated by genetic developments are already present in clinical settings but in nowhere near the intensity which will develop when the new knowledge turns into widespread new therapies. Stem cell research and human cloning are examples of issues which occupy politicians, journalists and the general public.

AIDS is another biomedical challenge shot through with ethical dilemmas. Like so many other diseases with which physicians battled over the years, waging an effective and aggressive campaign against AIDS requires attention both to its biological and bioethical dimensions. Sound strategy has from the start considered scientific and ethical dimensions of the disease. Commitments to find vaccines and therapies have been joined with campaigns to protect the human rights and dignity of people with HIV and AIDS. Efforts to stop transmission of the disease are combined with efforts to stop discrimination against disease bearers in employment, travel, housing, access to health care, and in hospital-based medical care provided by doctors and nurses.

AIDS, like the genome project, shows the inevitability of bioethics as well as the ever-expanding complexity of this field. On the pragmatic, concrete level there are the problems of confidentiality, allocation of resources, use of human subjects for research, public policy for schools, work place, prisons, and society at large, education and public campaigns, privacy, screening, informed consent, and on and on. No single aspect of the AIDS epidemic is devoid of its bioethical dimension.

Bioethics will continue to expand and to remain important throughout this century because biosciences will do both, and the two are inseparable. Institutional policies and codes and laws, national and international, political and professional, will have to be developed, then continually improved and updated. No end is in sight for the need of ethical professionals who are conversant with modern medicine.

CHANGES, CONTINUITIES AND CHALLENGES

a) The Early Culture

In the early days of the new discipline there was no sense of bioethicists exercising power either within the medical establishment or in the broader culture. The first practitioners came from different disciplines (theology, philosophy, law and medicine) and ordinarily understood their roles as theoretical clarifiers of problem areas or as creators of practical procedures for solving specific clinical problems. The early practitioners of the discipline divided along theoretical and practical lines. As far as I can remember, there was little interdisciplinary nastiness. Most of us who worked in academia taught courses in bioethics for graduate and undergraduate students, delivered papers at conferences on religion, philosophy and medicine, and offered consultation services to hospitals and hospital ethics committees. Complaints about the restrictive power of bioethics and bioethicists on medical practice and scientific freedom came later. Later still came suggestions that bioethicists were suspect of unethical behavior.

During the short history of this discipline things have changed. There have been many instances of political involvement on the part of bioethicists. Bioethicists who served on government committees or commissions sometimes were chosen in order to guarantee approval of some ethically controversial program or legislation. Something similar took place in the clinical setting. Some hospital administrators and/or local bishops "cleansed" ethics committees so that the committee's decisions would reflect a certain prescribed orthodoxy. Problems like abortion, physician-assisted suicide, reproductive technologies, embryo research, and cloning inevitably invited political involvement and divided bioethicists along ideological lines.

In the beginning different opinions were generally respected. The bioethicist was expected to be able to explain different sides of a debate fairly and then to develop a position without being disrespectful of opposing views. The early bioethicist had to take stands but did not have to advance an ideology. He or she would be expected to be able to state different sides of an issue fairly. Now, more and more bioethicists are advocates for certain positions and often engage in the same style polemics which one sees on TV shows, like *Crossfire*. Older bioethicists felt obliged to know their own background assumptions but not to defend ideologies. They did have to be sensitive to institutional missions and organizational priorities, but they were expected to be above being public relations staff for the organization. Most were more comfortable in the role of mediator than as a partisan or polemicist.

Bioethics in the beginning was different from other forms of applied ethics, and it remains so today. If the issue under consideration was exploitation of children, or forced abortion, or drug trade, the ethicists had to look carefully at facts and extenuating circumstances but ultimately had to take a strong stand against an evident evil. If the issue under consideration, however, is research with human subjects, then socio-cultural circumstances have to be considered. Final judgments on particular problems have to be nuanced. Medical research using human subjects is not intrinsically evil. New knowledge can serve both the research subjects and countless others. Without such research, medicine loses its scientific base. In clinical medicine, aggressive technological interventions on dying patients sometimes can be torture. At other times, with different patients, the same interventions can be life saving. Bioethicists have to be able to simplify complex situations, gather relevant data, develop clarifying concepts and apply appropriate laws and ethical standards. Bioethics then and now requires rigor in analysis and prudence in judgment.

Change in the discipline is necessary in order to address the challenges created by contemporary bioscience but it also has to preserve what bioethics started out being: a clarifying, mediating, discriminating, and critical voice in an essentially humane medical enterprise. Bioethicists have to avoid the simpleminded solutions of ideologues and pursue the development of sound, intercultural, international bioethical policies. In controversial areas they have to avoid the worst options.

In the beginning, bioethics tried to be helpful to professionals, and mediation was very often its principle role. However, there were times when the clinical bioethicists had to take a stand, had to say no, and possibly even had to anger those who asked for a consult. The discipline of bioethics at the beginning was not ideological, and yet neither was it ethically neutral. There were mainstream ethical stands dictated by considerations of facts, legal policies and guiding ethical principles. Early on, the discipline offered broadly agreed upon answers to frequently posed questions especially about the end of life. Bioethicists provided medical and nursing students whom they taught with ethical direction supported by law, philosophy and theology. Most early bioethicists also wrote newspaper articles defending or disputing or explaining the ethical dimensions of widely publicized legal cases.

b) Continuities

The beginning concerns gave early bioethics a casuistic style. Casuistry certainly has its limitations, but it also has its advantages. Bioethics today however has to be more and more international, because modern medicine and medical technologies both are. The clinical case can be the common starting point for bioethical reflection in any or every culture. And developing humane solutions to common clinical problems can easily be imagined to coalesce into ever more widely accepted standards. International policies can then gradually develop from similar resolutions of common recurring problems.

Besides being similar, cases can also be rather easy to solve because widely accepted principles directly apply. The test either for a bioethicist or a bioethics committee is to develop functional procedures and sophisticated patterns of reflection for handling tough cases. These, too, can and should become international. Relevant facts have to be identified; background conditions and underlying influences have to be made explicit; misconceptions and confusions have to be clarified. Clinical bioethics requires special sensitivities for the existential or lived dimension of an ethical problem. Moving toward defensible solutions starts with a sophisticated analysis of the subtleties of a case. Bioethics started with clinical problems, and bioethicists worldwide have to continue to improve their techniques for finding defensible solutions to complex clinical cases.

As important as it is, clinical bioethics can never be the whole of the discipline. Abstract and theoretical bioethics were present at the beginning and will always have a place. What is the good? How can justice be understood? Which principles prevail in conflict situations? These questions are always posed by clinical cases and should be addressed in the discipline of bioethics. Some bioethicists, however, should be concerned about certain new procedures and interventions even before they begin to generate particular problems in the clinic.

If bioethicists specializing in economic justice and access issues have to have demonstrated character traits and have to understand social sciences and the structures of health-care bureaucracies, the clinical bioethicist has to know medicine or nursing and understand the subtleties of the clinical context. Theoretical conceptual models ultimately have to be linked up with the existential particulars of a case. Reasoning for the clinical bioethicist moves from the practical to the theoretical. The test of a good clinical bioethicist is to have conceptual categories adequate to reorganize the data of a complex clinical reality, to understand it and to respond to it in an ethically defensible way. The clinical bioethicist does for confused and complicated cases what an insightful psychiatrist does for a confused and complicated patient. She figures out what is actually going on behind appearances and provides a number of possible options or explanations. He understands both the important facts and the less obvious background dimensions.

Court cases, legal statutes and government regulations have driven bioethical developments and given to bioethics a recognized importance. The fact that Joint Commission of the Accreditation of Health Care Organizations (JCAHO), which accredits hospitals, nursing homes, hospices and home health-care agencies in the United States, requires a mechanism for addressing ethical problems, continues to make bioethics important in U.S. clinical settings. License requirements put pressure on the discipline to establish programs to train ethics committee members in every type of health-care facility. This is true in the United States, and similar pressures in other countries are leading in the same direction. If bioethics will continue to be an important discipline in the 21st century, the next generations of bioethicists will have to know how the discipline began and its first stages of development. Changes are inevitable and continuity is essential for survival.

c) The Discipline Challenged

In a short period of 30 to 35 years the discipline of bioethics has moved from a private role of helping professionals faced with difficult clinical decisions to a more public role of advising legislators intent on constructing public policies about matters of health. But this was not the only development. Another evolutionary leap was revealed in a front-page story in the New York Times, August 2, 2001, with a headline which read, “Bioethicists Find Themselves The Ones Being Scrutinized.” From private mediation to public policy advising, and now to suspicion of being participants in the very practices which bioethics and bioethicists were meant to prevent.

Bioethicists now are being accused of striking financial bargains with powerful economic groups in order to provide the kind of ethical cover-ups which they want. The latest development is that bioethics has moved from a discipline of experts providing ethical advice to professionals and politicians to becoming a discipline suspect of a prostitution of ethical expertise. In just 30 years the discipline has moved from providing ethics experts for secular society (in roles similar to those filled by priests and theologians in more religious times) to producing ethically compromised power players in the world of big money.

It took no time at all for bioethicists to move from being trusted ethics experts for small groups of medical professionals to being self-appointed philosopher kings issuing ethical proclamations which promote their own financial interests, rather than what is objectively right or respectful of the rights of other persons. The idea of moral experts becoming morally compromised themselves is not all that rare. It happened to the theological moralists in more religious times, and now it is happening to their secular replacements.

Some bioethicists are employed by biotechnology companies and know what they are expected to do or to decide. Others work for universities (and presumably enjoy ivory tower protection), but they take money from corporations for programs and research projects. In either case the suspicion is created that bioethicists are permitting themselves and their discipline to be prostituted. This suspicion alone is sufficient to undermine a discipline which is the only source of ethical direction for our secular societies. Over the discipline's short history, bioethicists by and large have avoided consideration of the character and virtue aspects of ethics. Suddenly this dimension of bioethics and of bioethicists is front-page news.

Before the discipline of bioethics came into being, doctors and scientists did their own ethics. A medical ethics has existed since the very beginning of western medicine. In the 20th century the ethical problems multiplied. At the same time doctors and scientists became compromised by powerful economic interests. By so doing they destroyed the belief that they could be trusted to do their own ethics. Bioethics and bioethicists then stepped in. Now something similar is happening to the replacement ethicists after only 30 or so years.

Many of the first bioethicists were trained in theology or philosophy or medicine, and had already established reputations for being ethical persons. Now the number of bioethicists has exploded. Many of the newer bioethicists are persons who have graduated from a university program which did not require proven ethical character in the students, and may not even have connected the discipline with personal virtue and character issues. Many of the bioethicists who are most suspect of compromising their integrity by the ethical decisions which they make, don't even see the problem. They see nothing wrong with an ethics which they are doing for money or what is happening to the discipline because of their suspicious behavior.

Industries and HMOs hire only those bioethicists on whom they can count to support their interests. It may be too much to say that the company bioethicists are pawns or prostitutes, but this suspicion has been created. Practicing bioethicists have moved from being private mediators for professionals in clinical settings to becoming political power players, to becoming suspect of ethical compromise. The moving force behind this most recent development, of course, is money.

Those bioethicists who receive stock options or big salaries from their companies defend themselves by saying that they are working on the front-line in bioethics where the problems arise and where the action is. Those who criticize them or who work in less compromised settings are said to be out of touch with what is really going on.

However one feels about or judges this new situation, historically it is important to see how the discipline has developed over its short history and the circumstances or contexts in which a new generation of bioethicists are going to practice the discipline. If they work in a university setting, it is very likely that a corporation or a drug company will provide some kind of funding for a university conference or a university program and this creates suspicion. Bioethics journals receive money from industries and they might be offended by something said in a research report or in an article. Outside economic interests might even have an influence on what is accepted or rejected for publication. Big money has been known for some time to exercise control over the direction of scientific research and over what research results are published and what are not. Now big money is suspect of exercising control over what gets moral approval and what is morally disapproved.

Every discipline goes through changes. Some of the changes are healthy and contribute to the discipline's development. Other changes are not and threaten first the discipline's public image, then the discipline itself. If the present trends, temptations and compromises are not addressed, not just individual bioethicists will be suspect but the whole discipline of bioethics will loose its credibility. This is a challenge which the new generation will have to recognize and address. Keeping an eye on the history of the discipline should help.

THEORY

IV. Natural Law, History and Politics

The idea of an objective structure to human nature which serves as a foundation for ethical judgment is not unique to Catholic thinking. We find similar assumptions operating in most serious philosophical anthropologies. Certainly, the classical philosophers held the view that human nature and human reason are the ground for a universal ethics. The same is true, if not always overtly expressed, even in modern philosophy. Philosophers of human existence either implicitly (Nietzsche) or explicitly (Hegel) argue for the trans-historical validity of their philosophical views.

Jaspers is a good example of a thinker who is sensitive to the historical relativity of his images, and yet presumes to articulate insights which transcend his time and culture. He makes the point again and again that the image of human being which he proposes is comprehensible only in connection with the influences and events of his own lived experience. At the same time his philosophical discourse strained for a validity that reached beyond his particular personal or historical condition. He defended human reasoning about morality and what it means to be human no matter when or where the human being lives. He was trying to argue for what we call a Natural Law perspective.

Freud was not appreciably different from Jaspers in this respect. Critics have called attention to the strongly cultural and narrowly historical dimensions of Freud's reasoning but Freud himself intended to make a claim about the human condition no matter where or when it comes to be. Freud wanted his theoretical insights and moral standards to have universal validity and an impact on history beyond his age. Looked at from the perspective of psychohistorians, Freud intended to be a socio-ethical revolutionary. Marx is an all too obvious example of the same transhistorical objectives and claims. Movements in philosophy like Structuralism and Existentialism are rooted in particular experiences, and yet they too strive to articulate insights into the human condition and human morality which are universal.

To emphasize the inevitability of the historical/transhistorical polarity in philosophical anthropology we can look at a Catholic theory of human existence that grounds a clearly transhistorical ethics. Natural Law Theory, at least according to many of its interpreters, is universal in the sense that its basic vision and moral directives have validity beyond historical change. Catholic Natural Law presumes to be an articulation first of the nature of human reality and then of perennial ethical principles based on a universally valid anthropology. The basic philosophical vision and ethical principles of Catholic Natural Law claim universality.

Natural Law procedures and methodologies presumably can be used to arrive at ethical directives in ever changing historical and cultural contexts. This all-important objectivity about the human condition, however, sometimes is exaggerated so as to eliminate historical change in moral teachings. Sometimes, claims are exaggerations because human beings evolve, and so does human understanding and ethical judgment. Natural Law anthropology does ground universal ethical standards, but it also evolves. Natural Law thinking aspires to objectivity and universality and at the same time is open to the continuing influence of history and politics.

WHAT IS NATURAL LAW

Natural Law is not an ethical IBM computer. It does not provide automatic answers to questions of right and wrong. It is not even a law in the ordinary sense of the term. The modifier "natural" does not connote an object or a thing belonging to physical nature.

The Natural Law concept has its historical roots in Greek culture and philosophy. First, we see the concept expressed in theater, especially in the works of Sophocles (497-406 B.C.) In *Antigone*, for example, his main character insists upon her moral duty to bury her brother (Polyneices) even though the king (Creon) ordered that the body be left unburied. Which law prevails? Is it the will of the king which is the prevailing community law? Or is there a higher law reflected perhaps in traditional custom but grounded ultimately in an understanding of human being and the moral requirements founded on that understanding? Sophocles' point is that human nature and human reason are the ground of ethical duties which have to be recognized wherever human beings gather in community. Later, philosophers like Aristotle and Plato would argue explicitly that nature rather than convention is the foundation of both law and morality. Centuries later, St. Thomas would express this same notion by saying that if our natures were different, our moral obligations would be different. For over two thousand years, the greatest minds in Western culture agreed that there are universal laws based on human nature against which the laws of a particular king or ruler or legislature have to be judged.

Natural Law ethics begins with an attempt to work out, with the exclusive tools of human reason, the constituents of humanness: a defensible philosophical anthropology[33]. That anthropology has to take account of the fact that human beings are rational and free and must use both capacities to create an inner self. Natural Law Theory is based on a human nature and experience that is both rational and creative. Principles and laws which promote and protect human rationality, creativity and dignity are then derived from the philosophically articulated human structure. Socio-political conditions fall within the realm of the human and are therefore open to rational investigation and rational assessment just as individual personalistic phenomena are.

For reasons of preliminary clarification, we can speak of Natural Law as an attempt to construct a civic law and ethical principles from the interchange between a human capacity for freedom and rational reflection on one hand and the objective reality of human existence in its socio-political dimensions on the other. That which more often than not provides a powerful illumination of the ethical and legal derivatives from human existence is suffering. Our notion of what belongs to human nature and is constitutive of nature-driven moral and legal directives comes more often than not from contact with the absence of certain conditions. Suffering and injustice contribute to a vision both of the structure of human existence and of what we mean by humane law and ethics.

My objective is to concentrate on two aspects of Natural Law, its historicity and its broad social function. Focus on the historical character of Natural Law is a basic background constituent of a defensible Catholic Ethical perspective. It is required to balance an exaggerated rationalism which we often find embedded in official Catholic moral teachings. This is true of teachings about sexuality as well as about socio-political matters.

In the U.S. legal tradition certain human rights such as life, liberty, and pursuit of happiness are believed to be self-evident. What is proposed as self-evident in our founding documents reflects the classical Greek and Medieval Christian conviction that certain ethical principles are founded on human nature and are revealed by human reason alone. It is also true that the ethical principles/rights of the U.S. Constitution are the result of a long historical development. Natural Law arguments made by Catholic thinkers also introduce contingent, historical elements into their definitions of what are claimed to be self-evident, perennial principles.[34] Medieval Natural Law's notions of justice, for example, reflect the classical view that moral principles are universal. On the other hand, the content of the moral principles is shot through with presuppositions from the feudal social order.[35]

The moral principles or human rights declared in the U.S. Constitution have their historical content and origins in the Enlightenment and owe their existence to the hard work of certain founding fathers. The moral principles in the Universal Declaration of Rights adopted by the United Nations in 1948 have similar intellectual roots and had their own political advocates. Because the UN declaration is more recent, it is easier to identify persons like Eleanor Roosevelt without whom there would be no universal declaration of moral rights today.[36] Universal human rights presuppose a universal human nature which grounds the rights, but there is also an historical and political dimension to the rights declarations.

Natural Law thinking itself has a strong historical and political dimension. It is related to "the law" in the sense of the positive law by which society is ruled. Because human beings are social, their communities must be held together by laws which may be either good or bad. One type of bad law would legalize not the moral values based on human nature but demands from particular interests inimical to human dignity, justice, equality, etc. (For example, Creon's law that Polyneices not be accorded a decent burial, but rather be left to rot.)

The structure of human nature is far from self-evident, but certain nature-driven moral claims can be identified over long historical and political periods (the sanctity of life, justice, equality, truth). Human being shows itself to be structured, and a universal human structure is the foundation out of which value claims arise and universal principles are defended. A vision of human being and certain abstract moral principles based on that structure serve as standards for positive law norms for humane medical practice. Natural Law argues from an objectively given structure of human life to universal ethical principles and ideals. Good positive law gives concrete legal form to the universal norms and principles. Bad law suppresses or ignores them.[37]

Natural Law arguments are difficult to make in our post-modern intellectual environment. In a post-modern vision, there is no objective and structured human nature which is accessible to human reason and which can ground universal moral values. The only universal constituent of human being in the post-modern view is individual freedom. The individual person is not an example of a universal human nature, but a product of freedom. “I am whatever I want to be. And I can change whatever I have become.”

Philosophically, the post-modern vision of human being is one of a radical freedom best described as pure will, indeed, as will without limits. Consequently, there are no universal moral standards. The UN Declaration of Universal Rights or uniform moral standards for the entire world makes no sense in post-modernism. Each government, like each individual, does whatever it likes. Will and force is all that matters. When President George W. Bush declared that he would do whatever is in the best interest of the United States and not what is best for all human beings and the environment, he was providing a clear example of post-modern thinking.

Joined to a world system based on will and force is a tendency of individuals and nations alike to see themselves as victims. Paradoxically, individuals and nations turn out after all to have an objective structure imposed from outside: i.e., the structure of victimhood. In the incident of a Chinese fighter plane crashing with a U.S. spy plane, (March 2001), the discourse which followed was the discourse of victims. Each side claimed to be the victim. The worry created by the discourse was that either side would move toward world destruction based on its victimhood claim. The same claim of victimhood pervades the discourse of terrorism. The Terrorists justify killing of the innocent by the claim that they are victims. Victims whose behavior is ruled by will and force create a very dangerous world. It makes sense today to take a serious look at an older, more humane and more civilized philosophical and ethical perspective called Natural Law.

HISTORICITY IN NATURAL LAW

It was Gian Battista Vico (1668-1744) who first drew attention to the way human beings acted in history as the surest route to understanding human nature. The most reliable image of human being is known only through the history of cultural expression.[38] The most reliable image of human being is known only through the history of cultural expression.[39] John Courtney Murray, a liberal Catholic thinker and the most convincing contemporary exponent of Catholic Natural Law Theory, (see chapter 2) insisted as well that the Natural Law cannot be articulated independently of historical input. We must inquire into "the real man who grows in history amid changing conditions of social life, acquiring wisdom by the discipline of life itself, in many respects only gradually exploring the potentialities and dignities of his own nature."[40] Historical evolution brings to light new dimensions of in human nature which, according to Father Murray, struggle for expression and form. The Catholic perspective on Natural Law expressed by Father Murray amounts to a gradual historical development in our knowledge of human nature. Consequently, for him there is development and an historical dimension in all aspects of Natural Law morality.[41]

Historicity and evolution in Natural Law ethics is much more pervasive than conservative Catholic moralists and Church hierarchs have been willing to admit. Not only does our knowledge of human nature grow with historical experience, but human nature itself, by reason of its freedom and its capacity to intervene into nature with technology, develops and changes, requiring an ever-changing enunciation of the moral content of Natural Law principles. Conservative Catholic moral theologians and Church officials would have us believe that all Catholic moral teachings, even those most obviously culturally conditioned, are perennial and changeless.[42]

In what way does our knowledge of human nature as well as human nature itself develop? What does it mean to talk about evolving rights and principles of morality? Reason today can support without qualification both the inclusion of freedom as a human right and a proscription against killing other human beings. Historically, however, that same reason made a number of exceptions to this latter Natural Law principle. For many centuries in the West the killing of heretics and witches was looked upon as being altogether reasonable, indeed a holy thing. Today, only fanatic terrorists would justify such acts. We continue to kill criminals in the United States but this will surely appear to our descendants just as cruel and unjust as the burning of witches in Salem and the killing of heretics in post-Constantine Christianity. The society from which the criminal springs bears some responsibility. This altogether reasonable more modern Natural Law position developed gradually as humans developed in history, and it pushes toward constitutionalization in different societies. Natural Law-based positive laws and social policies, and Natural Law-based moral teachings in medicine both evolve, but the process is slow.

The developmental character of the Natural Law principle, "*Give to every man his due*" (*suum cuique tribuere*) offers us an example of Natural Law evolution and historicity. An assessment of what belongs to a person (*suum*) is ever changing, not only because of the changing conditions in which he/she lives, but also because humans change with their conditions. (*Yo soy yo y mis circumstancias*.) Private property has been defended in the Catholic social teachings as belonging to all human beings by Natural Law. This moral teaching is based on the supposition that property is acquired by human effort and is necessary for a decent human life. But the teaching is obviously historically conditioned.

This Natural Law moral teaching makes very good sense in an agricultural society and an underpopulated world. Does it continue to make sense when great masses of factory workers do not have access to property or the available property is already acquired? Suppose a community lives on an island, and after some time, all available property is privately owned. If private property belongs to every person who toils, then should not every worker have such property? If property is required for human beings to live a decent life, then those who came along after the limited amount of land was divided have as much right in the abstract as all others in the community. "Giving to every man what is his due" at this point in the historical development of an island community would require the dissolution of the privately-owned property in favor of some sort of socialization. Otherwise, increasingly large numbers of persons in the group would be deprived of rights. The principle of justice or equality does not change, but a development in social relations not only changes the content of justice and equality but also gradually contributes to the emergence of different human rights and of a somewhat different human being. We see this development in medicine with the enduring demand for access to health care and the different ways this form of justice is actualized.

In advanced technological societies, the function once performed by private property has been taken over by institutions like Social Security and state run health-care systems. As a result of historical development, both in human beings and in society, the concept of suum (what is due in justice) changes. The formal principle remains, but the content becomes different. It was often argued before the Soviet collapse that owning private property, which was a natural right in one period, might be a violation of that same right in another period. The right of all persons, in justice, to basic health care provided by the state survives even in socalist and non-socialist states.

Natural Law, therefore, is not a "given" code, permanently "out there" for all rational minds to see, but rather, as St. Thomas insisted, something to be discovered historically by a slow and torturous wrestling with an ever-changing reality.[43] Part of the ever-changing reality is human nature itself, and Natural Law must reflect these changes. As soon as we try to stop the process and declare that certain concrete moral directives are forever binding, we run the risk of placing Natural Law at the service of established interests. Natural Law would better be understood as a rational striving after an understanding of what is right and good but never fully achieved; a continuing search for moral meaning that has content but not in a finished form; a search for moral principles that have functions, but the functions sometimes are different.[44]

The good, the true and the just to which Natural Law points is an historical good, truth and justice conditioned by socio-cultural factors and tied to changing concrete situations. It is never a truth, a goodness, or a right, "given" in the sense of a static atemporal reality "out there" somewhere in the abstract and changeless world of eternal forms. It is, however, precisely this latter mistaken notion that has created a way of understanding Catholic moral teachings which encourages passive acceptance of past solutions and leads to the widespread impression that some official Church teachings are out of contact with reality. Only by recapturing a more modest vision can Natural Law ethics contribute to the on-going search for a dynamic, historical and objective but yet unrealized moral good. Only by recapturing an historical perspective on Natural Law can the Church provide convincing moral teaching about the complex issues with which modern medicine is engaged.

POLITICAL FUNCTIONS OF THE NATURAL LAW

A fleshed-out Natural Law concept cannot be arrived at independently of its functions, including political functions. Continuing the effort to develop a Natural Law Theory which is historical as well as objective and empirical, we must look into how Natural Law functioned in political history.[45] By showing how Natural Law actually functioned in history, we may gain a broader view of the Natural Law concept and see its relevance for the contemporary challenges created by new bio-technologies, the majority of which will require a legal and political response based on sound ethical principles and careful ethical analysis (e.g., cloning, genetic engineering, etc.) of the structure of reality.

1) In primitive societies, there was no real distinction between moral life, social life and the law. Consequently, there was no conceptual difference. Everything we understand by these categories was subsumed under custom or mores, including social usage, moral practices and lived precepts. There was just one reality, which Hegel called the simple ethical substance which was anterior to the development of subjectivity, individual conscience and law.[46] Both individual moral norms and juridical precepts were part of one ethical reality. Only when positive law, formally considered, was constituted as a separate entity written and promulgated were the conditions established for different functions of Natural Law Theory, first among which is the heuristic. Natural Law served to fill the gaps in the positive law, which was always the law of a certain people. Each such law had its *lagunae*, areas that were not covered in the written code. These gaps were filled by human reasoning which considered particular circumstance and general principles like justice. Today, presidential bioethics commissions provide an example of this function.

2) Law is primarily an institution of a particular people. When a community with its particular laws entered into contact with a foreign group and established trade and cultural relations, the second historical function of the Natural Law came into existence. The Natural Law idea served as a sort of international law to meet needs which particular codes of the related groups did not cover. Natural Law in the form of *Jus Gentium* performed a suppletory function; this time, however, not of gaps within a certain particular code, but gaps caused by relating one code to another. In this instance, Natural Law transcends the precinct of the positive law of a particular group to fill a very important international function.

This latter function was first directed to areas not covered in Roman law, but it reached its fullness in the Middle Ages and thereafter in the sixteenth and seventeenth centuries.[47] In the Roman period, *Jus Gentium* addressed those behaviors which were considered right or wrong despite customs of the particular cultures which made up the Roman Empire. The Middle Ages were governed by a positive legal code which amounted to a constitutionalized Christianity. In this latter context, *Jus Gentium*, directed itself to the problems centering around Jews and Arabs in medieval society, who had certain rights not based on membership in the Christian community.[48] In the sixteenth century, the same problem arose, this time, however, regarding the indigenous people of the newly discovered continent.[49] Legal relationships with dissident churches and between states with differing legal structures concerned Grotius who gave responses to these problems based on the Natural Law idea.[50] Historically, Natural Law filled the function of safeguarding the peaceful togetherness of different peoples, a function that in today's multicultural American society is a serious challenge.[51]

3) A third historical function stems from the fact that positive law is never a reality entirely enclosed within itself, but rather the projection of a *Weltanschauung* from which it derives its sense and to which, from time to time, it must appeal. The positive law is founded upon meta-juridical principles which are rooted in this *Weltanschauung*. The background principles of positive law, in the sense of the springs or roots from which particular concrete laws flow, historically go by the name of Natural Law. Every system of positive law is dependent upon a vision of the world which underlies and grounds a legal system. It is precisely this vision and its derived moral principles which allied military judges used as a base to condemn Nazi bioethical abuses at Nuremberg. The Nazi doctors and researchers were certainly not guilty of any violations of their own positive law code.[52] Certainly they could not have been judged by American positive laws. What the judges and later the United Nations did was to elevate meta-juridical principles of Western positive law systems to transcultural and universal status. The Nuremberg judgments and the Nuremberg Code which initiated contemporary bioethics were based upon a Natural Law presupposition.

4) The fourth and fifth historical functions of Natural Law arise on the occasion of a cultural crisis. When the generally accepted mores and precepts of a community lose their coherence and appeal, a crisis arises. Natural Law can and historically has, in such circumstances, begun to function in either a reactionary or a progressive way. In the first case (fourth function), the ways of the past (old laws) are considered natural (physis) as opposed to the new formulae (nomos) seen as products of a particular will imposed upon the people. The old law and old ways are looked upon as given and immutable. They are seen as laws of nature, an immutable Natural Law.

This same point of view and function is seen over and over again in Catholic moral teaching. In the papal commission called by John XXIII to review the Church's moral teachings about birth control, the commissioners by and large agreed that change was justifiable and right in light of changed circumstances. A few conservative theologians (and two cardinals who supported them) argued that change was impossible because the Church's older moral teachings based on Natural Law are both infallible and immutable.

5) The progressive (fifth) function arises out of the same historical situation. Reformers have always pointed to a vision of society which is more just than the existing one. (e.g., Martin Luther King, Jr.) They point, also, to a law that is more just and more rational, thereby invalidating the existing law and justifying its destruction. Here Natural Law, as opposed to the positive code in force, looks not to the past but to the future. The Stoics, for example, looked to an international society without slavery, founded on what they considered the Natural Law. The effectiveness of this function of the Natural Law concept is not the question.[53] Natural Law in the Stoic's viewpoint was utopian, minoritarian, individualistic and completely ineffective when confronted with the socio-economic conditions of a particular culture. It does, however, offer us an historical example of the fifth, progressive, future-oriented function of the Natural Law.

The period before and after the French Revolution provides us with more current examples of both the reactionary and the progressive functions of the Natural Law. The revolutionary moral claims of the French middle-class were considered Natural Law principles. Natural Law moral principles supported inalienable rights which were denied by the Monarchy, thereby depriving that regime and its laws of legitimacy.

These moral teachings gained positivicity when they became constitutionalized in the United States, French and other Enlightenment regimes. Eighteenth century Natural Law Theory was revolutionary and future-oriented. It provided us with a clear example of the anticipatory and pre-positive character of Natural Law, as well as its progressive orientation. Afterwards, during the nineteenth century, a reaction against the French revolution set in and again it was the Natural Law which was the rallying cry of moral philosophers and political theorists calling society back to the anterior juridical order (traditionalism with its notion of Law as arising historically and not invented by Enlightenment ideologies).[54]

The historical functions of Natural Law could all be understood as diverse ways of keeping positive law open. Natural Law opens up positive law (1) to the totality of culture (especially its meta-juridical background principles and ideals); (2) to the whole human society in the sense of the entire world (*gentes*) and thereby pushing toward a viable international law; (3) to history [a] to the past so that positive law will not become rootless, abstract, or closed in upon itself, [b] to the future in the sense of poignant positive law toward more progressive social possibilities rather than to situations dictated by existing power blocs.

NATURAL LAW THEORY FOR TODAY

The idea of Natural Law functioning in various ways is as important in Catholic moral teaching as it is in secular law and politics. In American culture, we are still imprisoned in the Kantian division between law (heteronomous and exterior) and morality (autonomous and interior). Legal positivism aggravates the situation by separating law from its socio-cultural roots and making it an entity functioning in splendid isolation from everything. To escape the distortion which accompanies either view, we need to recover the connection between law and morality.

Our secular American culture shows several different moral or ethical levels: (1) An autonomous and individual morality (insofar as such is possible); (2) Intimately connected with this is a cultural or community morality; (3) A heteronomous morality, usually religious (Jewish or Christian), but possibly scientific or humanistic; (4) A Natural Law functioning, as we have tried to show, as a quest for rationally defensible individual and social moral directives; (5) Positive laws (penal law, government regulations, rights legislation) which is constitutively moral; and (6) Positive law that is merely technical, but insofar as it establishes order (a value), is not completely devoid of moral content.

To this list, we can add a social morality which is an ethics inscribed in the very juridical and administrative institutions of the state. This is a concrete ethics as opposed to one that is abstract and purely academic. The discipline called social ethics is an important part of Catholic moral teachings, which addresses itself specifically to political institutions and economic entities. It crosses many of the established disciplines which study the social reality. The emphasis of social ethics is on the effects of public cultural structures on all human beings rather than on individual persons' intentions and personal feelings. As such, it could be termed neo-utilitarian. Evident beyond dispute is the fact that an individual or personal ethics, based on individual autonomy or good will, turns out to be ineffective for the solution of bioethical problems today, most of which have a strong social or legal dimension.

The Church and state today must both become more ethical realities: a Church and state not just of law but of justice and respect for human beings and human dignity. As a nation formed in the liberal tradition, we are a nation of law, but we are also a nation struggling to give more ethical substance to our social structures and institutions and laws. As a Church which is called to reflect the behavior of Jesus, we must give more ethical substance to our Church structures, procedural rules, our public institutional image and our institutional practices. (e.g., the way the Vatican Congregation for the Doctrine of the Faith investigates and disciplines liberal academics.)

This ethical level already exists in the sense that there are social structures inside and outside the Church and state with a built-in ethical content. This dimension on the moral landscape, however, did not just fall out of the sky. In the state, the positive ethical content of our social structures have been the result of gargantuan efforts by labor unions, civil rights groups and others, to inscribe ethical values into social structures and positive law. In the Church, whatever changes have occurred to make the Church's structures, laws, and practices more ethical have resulted from gargantuan efforts by Catholic reformers.

Few people today talk of the ethical dimensions of social realities in terms of Natural Law. Terminology like human rights, civil rights, rights to dissent, right of free speech, right to a decent wage, work-place protection rights, rights to compensation for injury, are more in vogue than the Natural Law language. As a matter of fact, however, this whole process of asserting and fighting for a more ethical reality in our ecclesiastical and secular institutions is a parallel to the eighteenth century revolutionary movements inspired by Natural Law concepts. In our time, as then, the drive is toward a concrete social ethic, concrete ethical laws and structures, and ultimately an ethical culture or community. Now, as then, we struggle with what we have described as the progressive or revolutionary function of Natural Law, the raising of ethical ideals to the level of positive juridical institutionalization. In bioethics, defensible ethical stands on most of the major issues have to be translated into law or at least have to influence law. Classical Natural Law Theory has a role to play in making contemporary medicine more humane and contemporary bioethics more convincing.

When Natural Law is mentioned, people tend to think of the Catholic Church and an objectively based ethical theory. As a sad matter of fact, however, today's Church has distinguished herself primarily as an exponent of Natural Law as it relates to individual issues in sexual morality.[55] Church authorities have been eloquent, unequivocal and very detailed in spelling out the individual obligations which they see flowing from Natural Law Theory as it applies to birth control and abortion. In instances in which Church administrators address social problems, frequently they become exponents of what we have spoken of as the fourth or reactionary function of Natural Law.

There was a time in the U.S. when Church leaders fought for a new social order with more justice and freedom for workers, the forgotten, and the oppressed. In Latin America, where the established order is often an established injustice and cruelty, Catholics who advocate change are often politically marginated both by higher secular and ecclesiastical powers (e.g., Liberation theologians and leaders of independent Church communities). Catholic thinkers have kept Natural Law Theory alive, and some moral theologians have employed the concept to push for needed reforms. The hierarchy, unfortunately, has frequently espoused a narrow version of this ethical tradition. Linking the term Natural Law to traditionalistic positions has been regrettable and explains its fall into disuse. There is, however, a continuing need to keep positive law open to rational reflection on historical, cultural, political, social and religious influences; and to the progressive/revolutionary futuristic functions of Natural Law.

Natural Law aspires to juridical positivicity. In itself, it is more moral than juridical, but potentially, intentionally and anticipatorily, it is law of the future, a prefiguration of a future juridical order. Its animating force, the source of its power, is not the individual who has been, with rare exception, incapable of doing anything to change social realities for the better. Natural Law, in its futuristic and progressive expressions, is a social force, incarnate in certain pressure groups and objectified in an ideology. This vision inspires groups that work for change. Painful as this is, it is often our moral embarrassment which most clearly reveals a needed new moral order. This is Natural Law Theory in operation.

Natural Law is a moral direction and exigency which is based on reason but points toward legal status. It prefigures social forms which are anticipatory of socio/juridical structures. Natural Law's direction toward juridical institutionalization culminates in its *Jus Gentium* function, with the juridical recognition of natural rights by all nations and all peoples. The movements toward European unity, drives for freedom of oppressed races and people, the UN declaration of rights, international courts to judge crimes against humanity are all realizations of the Natural Law in its progressive, futuristic, and *Jus Gentium* functions.

What has been accomplished up until now in the development of International Law has depended for its inspiration on unexpressed Natural Law presuppositions. International ethical standards in medicine point toward constitutionalization of a basic Natural Law vision and of Natural Law primary principles. In almost every case, principles like sanctity of life, justice and freedom create pressure to change inhuman and immoral practices. Arguments for these changes are based on reason and on what fulfillment of human nature demands. What reasoning means has been clarified by Thomas Aquinas and

John Courtney Murray, by Freud and Hegel and other formidable theoreticians. Contemporary appeals to reason and human rights or dignity have been made by any number of international organizations. All this suggests an implicit Natural Law reasoning that is radically historical and politically functional; a Natural Law which justifies, legalizes and strengthens certain objective, intelligible, universal moral principles.

NATURAL LAW AND BIOETHICS

There is an historical, cultural, political dimension to most bioethical problems. This is obvious and hardly a matter of great controversy. And yet there is a transhistorical, transcultural, transpolitical dimension that is not so obvious and is commonly denied by secular bioethicists. A Catholic bioethical perspective attempts to keep these two elements in dialogue. Conservative Catholic moralists try to offer quick and immediate solutions to complex problems based on traditional teachings which they claim are international and infallible. Secular moralists deny transcendent elements and reduce every issue to changing physical, cultural, historical components. Conservatives think that traditional teachings are right and unchangeable no matter what the new and changing circumstance. Universal values and transhistorical directives do exist, but they can guide ethical decision-making about particular issues in medicine only in dialogue with historical, cultural and political contexts.

Natural Law thinking supports transcultural and universal bioethical directives but employs them in continuing tension with particular contexts. To employ a baseball metaphor, a team uses universal rules, standards, directives and objectives, but they are always related to a particular field, this or that opponent, these particular conditions. In bioethics, the principles rooted in the universal structure of human life have to provide direction and regulations on the playing field of contemporary life and medicine. In developed countries, the playing field is high-tech, unequal, full of class discrimination. In developing countries, the playing field is permeated by poverty, malnutrition and unavailable therapeutic medicines. What do the objective Natural Law principles of justice and equality require? The messages in the sense of particular right or wrong judgments are different in different contexts. In all contexts or on all playing fields the principles operate, but the way they play out is very much influenced by the different teams and different fields.

Equality, justice, freedom, truth, integrity, beneficence, caring, all these universal ethical values are involved in the bioethics game. But the particular moves which they generate, the needs which they address, and the conclusions to which they point are different. Health care has to be just and beneficent, caring and respectful of autonomy because these are required for human dignity, but what these principles dictate are very different on different playing fields. The universal Natural Law principles will require very different particular moves depending on the other team and the field on which the game is played. A liberal Catholic perspective in bioethics tries to keep in play the universal and the particular aspects of Natural Law reasoning.

V. Natural Law, And Universal Standards

INTRODUCTION

The influence of Natural Law in history and politics has to be recognized and then the Natural Law perspective has to be related to concepts like universal human rights in medical matters. What we know to be right and good in medicine is frequently violated, and abuses are difficult to remedy either because there are no universal laws or because no mechanism exists legally for applying transcultural ethical standards. The Natural Law perspective, for example, shows us that it is wrong for doctors to abuse patients and yet often they do so. Why? For economic reasons, for fear of job loss, for political obedience. In some cultures, doctors cut off hands and arms and ears because they are told to do so by military and political forces. Here universal ethical standards conflict with particular powers and call for stronger legislation of universal rights and standards. Many Natural Law-based bioethical norms need to be legalized so that they can overcome immoral political forces. International bioethics tribunals would be one way to force regimes everywhere to take basic human rights and bioethical principles seriously.

Violations of universal bioethical standards are not things of the past. We know how closely doctors and medicine were linked with military and political powerssuch as the Nazi state. This linkage made doctors strong advocates of the state and then instruments for carrying out thoroughly unethical research on human subjects. Psychiatrists were particularly vulnerable. They often carried out painful procedures on political dissidents with powerful medications, and they reported on their patients to government officials. Some of these abuses still go on.

In some political systems, dissidents are excluded from health care and doctors carry out immoral measures. Today in the United States, the hunger for health data is insatiable and doctors often violate confidentiality to provide the data. Military physicians are often in the service of the state and not the patient. The same is true of managed-care physicians. If doctors in the U.S. and other countries are working for the government, for insurance companies, for HMOs, they are in the same relationship with non-patient interests that Nazi and Marxist physicians were during World War II.

The situation of medicine worldwide calls out for a Natural Law perspective, for objective, intelligible, universal ethical standards, and for the raising of transcultural rights or values to international legal status. Why can't we have universal bioethical laws against torture, research subject abuse and female circumcision?

The answer is, we can. Legal experts from several countries and several international organizations have recently created a set of international legal standards for identifying and judging criminal behavior. The experts developed ten principles for identifying international crimes and guidelines for conducting international trials of those who perpetrate crimes against humanity.[56] What has been created is a universal jurisdiction for international crimes. But what has been created could serve as a model for something similar focused on medical practice.

Doctors everywhere are culturally conditioned and consequently awful ethical violations sometimes are ignored or repressed. More influence on the ethical behavior of medical professionals everywhere from legalized universal standards is needed. Natural Law Theory and concrete moral legislation belong together as much in medicine today as in any other field. Lawmakers have to know about the issues and have to make laws which protect the sick and vulnerable, human life. International laws covering medical practice have to be enforced and violations have to be punished.

The chapters on technology below look at threats to human life and to the whole biosphere. How to respond to these threats is the question. Can legal limits be established based on a Natural Law vision? There is no doubt that legally established limits are critical for survival of the planet. And there is no doubt that legislation requires philosophical and ethical justification. A Natural Law perspective can meet this challenge.

Human beings seem to have a tendency to reduce deep and elusive complexities about human existence to short, simple, slogan-like formulae. Abortion complexities, for example, generate oversimplified solutions contained in slogans about life and choice. The same is true of pressing political and religious issues. Slogans, in their turn, reinforce artificial group boundaries formed around these synthesized belief statements. Most persons belong to some such group: a protest group, an advocacy organization, a political party, a sports club, a professional association, an ethnic unit, a civic organization, a gang, a nationalistic or military association. Sect-like groups reinforce the simplified belief statements, strengthen unfounded superiority claims, and create separate cultures with their own values.

Bolstered by fellow believers in the group, along with visual identifiers like badges, metals, name brand clothes, styles, hats, etc., the endurance of the artificial social boundaries and separate cultures is guaranteed. Each group becomes something like a church or, more precisely, a sect. Each sect adopts some version of "we are the only true church" idea. Those outside the petty cultural boundaries are considered lost, enemies, odd, different or not worthy of respect.

Societies, in both developed and developing nations, are becoming fragmented into separate cultures of fellow believers on the one hand and moral strangers on the other. People feel the need to belong to one or another group. Some people need to belong among "the saved" or "the superior." They split off from "the others" and make few attempts to communicate outside their sectarian "communities." Consequently, the separate cultures become stronger while the broader culture, or civil society, weakens. When civil society disintegrates, prejudice, racism, sectarianism, flourish. Some of the basic ethical values and universal human rights have become major challenges as we start the third millennium: religious tolerance, respect for persons who are different, protection of patients and research subjects.

How can something so basic to human survival as respect for different human life forms become so easily endangered? What happened to the universal rights? What happened to that sense of wonder and respect for all humans which is grounded upon looking into another person's face? Who can come into contact with a small child without feeling tenderness and love? Sad to say, the answer to this last question is that there are many people who lack just such tenderness and love if the child or the other person is from a different racial or cultural or religious group. (Osama bin Laden takes the life of innocent women and children in perfect tranquility because they are "infidels".) Even for many practicing Christians, the commitment to respect for all others has been lost. The revolutionary dimension of Natural Law as the foundation of universal human rights is ignored. Only members of their tribe are considered right and worthy of respect.[57]

It seems appropriate then, once again, to raise the issue of a Natural Law perspective and to inquire about the possibility of recovering the foundation of universal human rights. We need to recover universal values and articulate universal ethical standards: ones which would draw us into respectful, indeed into friendly communication with one another; ones which help us to appreciate our groups or cultures without losing appreciation for others and ones which set down legal standards to protect these values. Is it possible that the experience of illness and an ethics of medicine might provide the very help we need to find our way back to a universal standard of medical ethics? If so, a Natural Law-based medical ethics will have to mount a convincing challenge to discriminatory cultures and sectarian definitions of right and wrong.

CULTURE-BASED MORALITY

Different cultures are assumed to have different moral values. Stated in its usual way, people are different and so too are their morals. The United Nation's Universal Declaration of Human Rights had to overcome stiff cultural opposition. Particular cultures in post-modernism are taken to be the sole source of ethical standards, and the term "culture" refers to ethnic, racial, religious, linguistic, nationalistic divisions, with their beliefs, values and behavioral patterns. The assumption is that no way exists to ground an objective, rational, universal ethics. We are assumed to be hopelessly divided and morally justified in pursuing particular individual and cultural interests. A person might be pushed to admit that babies and children are the same everywhere, but the idea that common beginnings might be the ground of a common human structure from which common ethical values may be derived tends not to be given serious consideration. Universal human traits, if they exist, are assumed to disappear in the course of human development and consequently only culturally different people remain.

Without universal rights, values and standards, ethical disagreements have to be settled either by force or negotiation. Radical cultural relativity logically leads either to pessimism about objective moral standards or to an ethics of negotiated permission. Cultural groups in the U.S. struggle with one another politically, not to advance tolerance, but in order to impose what they consider to be right. Morally divided groups fight to legislate their particular ethical convictions.

Right-wing American politicians, along with militia members and hate groups, speak very negatively of the UN and campaign against any suggestions that the U.S. might be bound to respect international human rights which are at odds with what they believe to be national interests. What is right and good for one group or nation may indeed be disastrous for other people, but it is taken for granted that each must push and fight to impose its own values and interests. The idea, for example, that anyone should sacrifice national or ethnic or economic advantage for what is good for people in other cultures or good for humankind is considered to be political heresy and anti-American.[58] The postmodern assumption is that universal ethical obligations are nonexistent. The goods of particular groups or cultures are all there are. When these come into conflict, there are only two options: either to fight to impose one's values on others, or to negotiate a climate of minimal tolerance in which every subculture is on its own, in a free, competitive market environment. The options are either all-out war or economic competition in a global marketplace.

GETTING BEYOND CULTURAL DIFFERENCES VIA MEDICAL ETHICS

The proposition that human beings, so divided culturally (and religiously), can come to an agreement about a universal anthropology from which universal ethical standards could be derived is admittedly a difficult one to argue.[59] But the idea that universal medical ethical values and standards might be derived from a common-sense understanding of the ends and purposes of medicine at face value seems more feasible. If babies are the same everywhere and solicit from emotionally-normal people a disposition to protect and help, the same is true of people with illness and disease. Cholera, TB, malaria and AIDS create the same painful and needful human conditions everywhere. These commonalties can serve as a basis for developing standards for how patients everywhere should be treated. Said differently, the needs of sick people can ground an objective and more humane medicine which can become the foundation for universal ethical standards.

The possibility of developing objective ethical norms based on rationaly accessable universal medical conditions is enhanced by the fact that in modern medicine, disease is understood by using universal scientific categories. Mainline medicine's response to disease is based on research, i.e., laboratory science's universal methodological procedures. Scientific medical research ultimately requires the involvement of human subjects. In 1947, experience with the way Nazi doctors used human subjects for scientific medical research led to the first secular expression of a universal medical ethics. The Nuremberg Code's ethical standards have been expanded and refined, but they remain in place everywhere as an example of an objectively based universal ethics.[60] No matter where scientific research is carried out, it creates the same ethical dangers both for scientists and subjects.

Medical technology shapes both scientific research and clinical treatment of disease. How can technology be used humanely on people who are ill? How can inhumane use of medical technology be avoided? How can vulnerable patients' needs best be addressed when the available technological interventions are dangerous? These questions are asked wherever medicine is practiced. It is easier to argue against inhumane uses of medical technology than it is to legislate against inhumanities of other sorts. Could anyone justify the way Nazi doctors used dangerous technology on uninformed patients? A culturally based ethical relativism seems less obvious once one enters the world of scientific medicine.[61]

Modern scientific medicine, in effect, is able to generate a medical ethics which transcends particular cultures, just as illness, disease and scientific research do. One year after Nuremberg, the U.N. Universal Declaration of Human Rights was promulgated.[62] Revulsion caused by misuse of human subjects generated the same ethical standards for proper treatment of research subjects everywhere. Later, the standards governing the use of research subjects were applied to the treatment of patients.[63] Cultural standards certainly exist in modern medicine, but so too do universal ones. Within the smaller world of medicine it is easier to recognize ethical standards which are peculiar to cultures as well as those which transcend cultures. Now the challenge is to evolve and apply the universal standards and at the same time to show respect for cultural values.

The international bioethics standards will be familiar to most readers. Patients and research subjects, for example, must be treated with respect, and respect involves some version of what we call informed and free consent. This ethical requirement remains valid even in cultures where women have little power to control most aspects of their lives. This first legalized ethical requirement enunciated at Nuremberg was followed by many others. All forms of physical torture with physician involvements, for example, have been condemned. The same is true of the use of psychiatric interventions for punishing political dissidents. Ethical standards proscribing psychiatric intervention on political dissidents were issued by the World Psychiatric Association despite opposition from the countries and cultures which habitually employed such practices.[64]

Modern medical ethics, in effect, can serve as an encouraging example of international campaigns directed toward creating universal ethical policies and concrete ethical standards. Concrete policies implement abstract values like respect and justice which are generated by a Natural Law perspective on the human condition. Today, universal medical ethics actually exists and is expanding. It is an ethics grounded on objective medical realities and with an unarticulated Natural Law supposition. In medical ethics, one culture does not impose its values or ethical practices on others. Rather a dialogue occurs about how the universal and transcultural principles will be socio-politically implemented. National ethical review committees are usually required to apply the universal standards with sensitivity, but the universal standards based on the nature of research and clinical medicine are not set aside or subjugated to contrary cultural custom.[65] Natural Law-based goods and values point toward legalization, both nationally and internationally.

To say that different cultures generate different ethical standards is a platitude. It ignores a more substantial truth, i.e., that trans-cultural commonalties exist because we share a common human nature and we face a similarly structured objective reality which can be understood and can be the ground for ethical values.[66] Admittedly, universal ethical standards are difficult to recognize and to formulate. Different geographies, histories, languages, religions, racial and ethnic strains obviously exist and so does the obvious conclusion that these contribute to different moral practices. But different cultures also have common elements derived from the objective realities which all human beings face. Ethical commonalities exist based on those shared human conditions, and they call for legislation and socio-political implementation.

The content of any culture changes and develops. Evolution is as clear in ethics rooted in cultures as it is in other areas of life. Ethical values and standards, even though at present divergent, can move toward convergence under the influence of shared experience, especially objectively grounded experience like sickness, medical treatment, and human involvement in research. Modern medicine, in both its research and treatment modalities, pushes different cultures toward ethical convergence and transcultural ethical standards.

THE CONTENT OF A UNIVERSAL ETHICS

Arguing against culture-based ethical relativism is made difficult by the fact that all important ethical commonalties are embedded in reality but hidden, instead of being on the surface to be perceived and easily articulated. Common ethical values embedded in the same human nature and the same objective medical conditions often are expressed differently. North Americans, for example, prefer rationalistic, secular, principle-based talk about values. Another culture may express the same ethical insights in religious language or in virtue categories, in more narrative and less rationalistic styles. But these differences do not destroy the underlying foundation of a Natural Law-based ethics which obliges researchers and subjects, doctors and patients, no matter what their cultural identities and religious beliefs.

An ethnocentric ethics may be more easily expressed, but difficulties in expressing universal medical ethical standards and constructing universal jurisdictions are not insurmountable. We have seen many instances of universal standards being articulated by the United Nations and other international organizations. Recently we have seen the publication of fourteen Principles on Universal Jurisdiction for crimes against humanity. Universal laws and universal jurisdictions for applying the laws then are not unique to medicine. Legal experts from different countries developed the Principles on Universal Jurisdiction to govern trials for crimes against humanity and disputes between governments. The trials of prominent figures like Pinochet and Milosevic call attention to this new dimension of a universal ethics.

Internationally, we already agree that medical treatment has to be beneficial to the patient, or at least it must strive to be so. Treatments as well as clinical experiments usually involve some risk which has to be measured against anticipated benefit. Patients in different cultures can certainly respond differently to what medicine has to offer, but balancing risks and benefits to the patient is a trans-cultural requirement. Harming a patient, or merely feigning help, or refusing to help is wrong everywhere in the world. These ethical values have been legalized and call for continued and extended judicial application. They are given juridical status today in international declarations.

In the past, these standards received similar status in medical ethics codes: Hippocratic, Chinese, Persian, Indian, Hebrew, and Japanese. The same ethical standards were articulated at different historical times and in different cultures, and they provide powerful examples of objectively based Natural Law values. The ancient codes all called for physicians to suppress self-interest in favor of the interests of the sick person. Altruism toward sick and needy patients was always an ethical obligation. This special form of love broke down into universal proscriptions against killing and harming and taking sexual advantage of patients. Historically and transculturally, it also meant guarding patient's secrets and confidences.

Today we speak of respecting patients by providing them with information about their condition and complying with their choices from among real medical options. This respect may not be as individualistically expressed elsewhere as it is in a United States-style of informed consent but its absence cannot be tolerated no matter what a particular cultural custom might dictate. Patient respect in the form of a requirement of informed consent is a modern addition to the older values of beneficence and non-maleficence. It is, however, equally universal and another example of how Natural Law-based medical ethics, like all forms of ethics, evolves.

It may sound strange to some readers, but historically, truth was not a medical value. In fact, doctors were considered exempt from the requirement to speak the truth. Today things are changing in this regard. There are still exceptions to the requirement of full truthful disclosure because subtleties exist in determining just what truthful communication in medicine involves. But a requirement to tell the truth, or not to lie, has become another transcultural ethical standard. Without the universal value of honesty, relationships cannot develop among human beings, and this is especially true of therapeutic relationships. Lying is an enemy of curing no matter what the culture. Manipulations of patients through lies pollutes the doctor patient relationship and the whole context of medicine. Universal medical ethics requires of doctors that they struggle against self-promotion because this easily leads to a compromising of truthfulness. Lying and other forms of willful patient deception are violations of a transcultural medical ethics. What is true of the doctor/patient relationship is even more true of the researcher/subject relationship. Without truthfulness, medical research involving human subjects is unethical.

Paradoxically, respecting different cultures is another important universal ethical value because it is tied to respecting human persons who are what they are, not totally, but to some extent at least, as a result of their culture's language, art, literature, customs, religion and law. Respecting a culture involves respecting the identity of persons formed in that culture. Violating some cultural values and practices amounts to violating persons. When conflicts develop in a mixed-cultural doctor patient relationship, conceptual clarity and careful procedures for working toward morally defensible resolutions are one more transcultural ethical requirement. Mediating procedures are especially important in situations of apparent conflict between universal standards and particular cultural norms.

Respect for culture is a universal but not an absolute value. There are limits to the respect due to cultural practices. Cultures cannot ethically treat sick people any way they like, cannot use medical technology any way they like, cannot use human beings in research any way they like, cannot intervene into the human genome any way they like, cannot make whatever laws they like to control population. The issues raised by modern medical practice may generate different cultural responses, but no response is ethically admissible just because it is a cultural practice. Certain basic values based on reason's insight into human nature and medical practice requires legalization even when opposed by a culture.

What one culture approves may create unacceptable impositions on people in another culture. A rich nation, for example, which approves the purchase of organs would create a terrible imposition on a poorer neighbor pushed to self-mutilation in order to survive. If money alone is allowed to determine access to treatments, only the most wealthy will live, and they will do so at the expense of the most deprived. That would be wrong even if it received approval from free market capitalist cultures. Reproductive health practices also may differ from culture to culture. The same is true of confidentiality standards and what is considered to be equitable health-care delivery. But this does not mean that anything goes or that no limits exist to what one culture may approve or disapprove.

If certain medical ethical standards are transcultural, so are certain medical ethical dilemmas. How much autonomy should the patient have? How is this balanced with a physician's professional obligation to do what is beneficial? How much power does the public health officer have? How is this balanced with individual patient autonomy? Medical practice is an economic reality, but how can economics be kept from turning medicine into a purely monetary enterprise? Research using human subjects is necessary, but at what point does it become manipulation and misuse of human beings? Technology inevitably plays a role in human reproduction, but what are the limits of technologically engineered reproduction? (This issue will be discussed in chapters on technology.)

Cultural differences do not require ethical relativism but rather fuel a drive to articulate standards which oblige us all. A universal medical ethics is based on common dimensions of human persons, common scientific assumptions, common conditions created by disease, and the commonalties inherent in the relationship between a sick person and the doctor from whom help is sought. Given the shared background assumptions of modern medical science and the shared technologies of modern medicine, humane medical help can and should conform to common ethical standards no matter what the cultural context. Medical ethics can be transcultural because the science of medicine and the experience of sickness are both universal. Universal ethical standards may not be obvious, but intercultural dialogue to identify them and to articulate them is a worthwhile enterprise.

UNIVERSAL PRINCIPLES AND CONCRETE NORMS

To argue convincingly for the existence of a universal medical ethics is the important first step in a complex larger project. The next step is to attempt to move from abstract universal rights and values to concrete norms and rules covering specific situations. Abstract values and principles are helpful in determining what to do in a clinical situation, but something more is needed.

It is one thing successfully to ground a universal medical ethics on a structured objective reality which is expressed in the form of basic Natural Law principles (respect, beneficence, truth, love, life, justice). It is another thing to apply these objectively based values coherently to concrete practices and to develop ethical policies for particular circumstances. A workable universal medical ethics for a more humane medicine must be applicable to every day circumstances of medical practice in any culture.

The first thing to note about moral choices made in actual medical practice is that they tend to be made with some sense of urgency. They are not the kind of choices one can afford to mull over or talk about forever. And yet they are full of subtleties and complexities. The choices to be made are important both for the patient and the doctor. The pressing urgency of many clinical decisions, however, cannot justify instinct-driven practices or shoddy decision-making procedures.

What has been argued for thus far is a medical ethics which one finds in international medical declarations, codes and conventions. It tends to be an ethics of very general standards. Medical ethical codes usually speak about ethics at an abstract level of discourse. The Convention on Human Rights and Bio-medicine of the Council of Europe[67] speaks of dignity of all human beings, integrity, equitable access, therapeutic benefit, the primacy of human being, free and informed consent. The consent requirement is broken down into standards for adequate information and free consent as far as possible, even in emergencies and even for mentally disabled patients. One specific ethical policy was formulated against the disposal of human body parts for financial gain.

Work on this Convention began in 1989 and concluded with final approval in November 1996. Subsequent policies intended to serve as international standards evolved from the 1948 United Nations Universal Declaration of Human Rights and subsequent international conventions and covenants. The Convention points toward creating an intercultural agreement about ethical standards for biology and biomedicine.[68] But making ethical values juridical is not an easy task. It is what most revolutions are all about.

Expressions of universal standards assume that in different cultures these will be somewhat differently applied. In fact, in every instance, a gap will exist between the universal ethical standards and the cultural circumstances and medical contexts in which they will be applied. Universal standards alone are never enough. These must be made concrete in laws and policies. Before they can generate ethical policies for particular cultures, a detailed examination of cultural circumstances and clinical contexts is required. A flat-footed, direct and unsubtle application of ethical values will oftentimes create more harm than good. Said differently, universal ethical values and principles are fundamental and yet will usually require some adjustment in order to be adequately applied or implemented. Culture will never invalidate universal ethical principles or require that they be violated. The customs of a culture, however, will always require consideration for the way principles are applied via rules and norms.

THE CASE OF FEMALE CIRCUMCISION

Female circumcision is a term which describes different types of genital surgeries performed usually on very young girls in many African, Indian, Malaysian and Middle Eastern Cultures. It involves partial or total clitoridectomy, sometimes with an added surgical closure of the vaginal opening. An estimated 100 million women are subjected to such procedures which are usually performed without anesthesia and in non-sterile conditions. Besides the immediate associated pain, infections are common. So are long-term ill effects like infertility, painful intercourse and diminished sexual pleasure.

Female circumcision provides an example of the relationship between abstract universal principles and concrete cultural contexts. Some who deny the validity or even the possibility of a universal medical ethics opt simplistically for cultural relativism. They accept the way certain cultures mutilate the sexual organs of young women. Whatever is right or acceptable in a particular culture will be considered right and acceptable. General principles like beneficence, non-maleficence, truth, and respect are simply set aside. On the other side are universalists who proceed deductively from the abstract principles to immediate proscription of any and every practice related to female sexuality. Cultural contexts are left completely out of consideration. A liberal Catholic middle-ground ethical perspective gives close and careful attention to cultural contexts in order to understand just what universal values and principles require and how they should be applied to change the situation of young girls whose loving parents believe that female circumcision is a necessary good.

In most cultures where it is practiced, female circumcision is a "woman's thing," i.e., controlled by female rather than male family members. It is such a common practice that it is understood by many to be "normal" and "natural." In addition, it is considered a requirement for marriage, an ethnic marker, a way of proving virginity and protecting family honor, a way of enhancing a husband's pleasure. Religious support for the practice in Islam is usually linked to beliefs about ritual purity and the need to control female sexuality rather than to the authority of Islamic scripture. (Male circumcision on the other hand is required both in Islamic and Jewish scripture).

Because female circumcision is strongly rooted in certain cultures, it is even practiced on non-Muslim girls. It is a "marriageability" requirement for them, too, as well as a way of keeping their husbands sexually satisfied. The daughters of more affluent secular parents may be sent to physicians in order to reduce pain and infection. But even for them it is believed to be a protection from sexual involvement when young ladies have to wait until after their university educations in order to marry.

Objections to female circumcision based on universal ethical principles are strong and spreading. They can be traced centuries into the past. Catholic missionaries at first considered the practice to be wrong and forbade it, only to relent when Catholic girls found themselves unmarriable. More recently, feminists have embraced the anti-circumcision cause and have involved the World Health Organization in a crusade which is carried out under the banner of Women and Children's Health. Female circumcision practices have been publicized in the popular press and are labeled as mutilation, torture, barbarism and ritualized abuse. Because this type of circumcision is practiced even in Western Countries by immigrants from Africa and the Middle East, legislation has been introduced both in Europe and North America to ban the practice. In Canada, the College of Physicians and Surgeons developed a policy statement which bars the procedure. By contrast, the vast majority of women in the cultures where female circumcision is practiced still consider it a normal preparation for womanhood.

To recognize the complexities involved in applying universal ethical standards to cultural practices is not to give up on the project. An effective application of an international ethics requires subtle steps toward concrete application of broad principles to particular health-related practices. Sensitive and nuanced application begins with a thorough understanding of how the health-related practices are perceived by persons within a culture. Outsiders don't always get a clear picture simply by looking at a practice through a Western cultural lens.

Patience is required. Cultural practices in developing nations may not change as rapidly as they do in modern societies. To extract a cultural practice from its settings in order to alter it, is to oversimplify a complex and essentially developing phenomenon. One recognizable instance of this evolutionary dimension is a widespread acceptance today of western scientific medical interventions in life-threatening situations. Trusted local medical practitioners can help to avoid oversimplified misunderstandings of a culture and insensitive applications of the universal principles. More humane doctors can start up dialogue, carry out mediation, and finally effect change.

The aim of applying universal principles to problematic cultural practices is to prevent a health-related harm and to advance medical good. In order to accomplish one or the other objective, the consequences of the cultural practices and required changes must be carefully assessed. Sometimes harm is done by judgmental moral crusaders who come over as self-righteous and exemplify cultural hegemony.

Finding the right voice for speaking about humane medicine is critical. Questions like how much can be said and to whom, have to be addressed. Simple condemnations may be clear and honest but seldom accomplish the intended good. Statements that express understanding of how questionable cultural practices developed historically can modify and moderate dialogue about change. Success depends upon finding the right language for expressing the moral judgments and the right sources of moral authority.

If moral judgments are being made by persons from another culture, some language which recognizes the moral inadequacies of both cultures usually helps. This reduces the possibility of an ethical judgment coming over as a form of moral imperialism. Some statements of respect for the culture whose practices are being judged always help. An admission and recognition of the perspective from which a critical judgment is being made is both honest and helpful. For example, "from the perspective of modern western medicine, or according to universal ethical standards, this or that practice causes serious and long-term physical damage. It aggravates already existing symptoms and provides no compensating medical benefit." It is also important for professional associations to support and protect nurses and physicians from authoritarian regimes when they speak out against the practices which violate universal standards.

CONCLUSION

In a Natural Law perspective, there is no way of avoiding the ethical dimensions of human behavior. Every research scientist, for example, whose research will affect the reality which human beings confront has to ask himself or herself whether it is right or good to be involved in a particular project (e.g., gene mapping or reproductive engineering or nuclear physics). If this is true of work in physics and genetics, it is certainly true of work in an applied science like medicine. Every doctor, in every culture, with every patient, has to ask whether what he or she proposes to do is ethically right. This is true because human beings everywhere have a common structure and so too does the reality which they confront. It is more than understandable that historically, healers in every culture developed similar ethical codes. Sensitivity toward the needy ill was always required, as well as respect for human dignity, a guarding of secrets, and a commitment not to take advantage of patient vulnerability or to do patients harm. At the level of general principles, an international or global medical ethics obviously exists. Natural Law provides a coherent and convincing explanation of this. Making general principles into concrete laws or universal medical rules is the challenge.

Consideration of culture in the application of a universal medical ethics is important but it can never equate to a betrayal of the universal standards. To keep this from happening, doctors and nurses, who already have a certain cultural formation, need to be formed as well by the universal ethics. It is all too easy for health-care professionals to be so influenced by cultural/political interests that they violate historical and universal standards. Universal norms and policies (for example, against torture, abuse, mutilation) have to be given juridical and professional force in order to counter cultural and political pressure on medical professionals who earn their living working for others (e.g., a government or the military). International professional associations need real political power to counter the pressure coming from within a culture. Doctors and nurses in every culture have to receive ethical education and must experience the solidarity of the universal medical community in order to stand up against pressures to violate ethical standards. Unless a global or universal medical ethics is taught effectively to future doctors and nurses, it will never be translated into practice. And unless professional medical associations monitor medical practice and respond to reports of ethical violations, the universal ethics will amount only to a list of platitudes. If ethical violations occur, global medical associations have to act. Ethical reforms can be carried out only by powerful groups.

The medical ethical failures which took place in Nazi Germany provide lessons which cannot be ignored. In the early 20th century, German medicine was the most highly regarded scientific medicine in the world. But doctors who worked for the state and/or for the military first labeled political dissent a disease and then reported on their patients. These initial ethical violations were followed by closer cooperation with political powers and then worse violations of "inferior persons" (mentally ill, Gypsies, Jews). First, they called for less treatment and ended up calling for mass murder. The apparent ease with which many doctors and nurses carried out state-sanctioned immoralities shows how important it is to provide explicit and thorough ethical training for medical practitioners. Then, a strong awareness that medical behaviors will be monitored by international medical associations is necessary and there has to be some legal system for bringing violators of international standards to justice. Any use of medical personnel or medical interventions for accomplishing non-medical objectives (for example: repression of dissent, military readiness, preference for male babies, insurance eligibility) has to send an immediate signal to practitioners and must be resisted.[69]

Some less than ideal health-related practices must be allowed to stand, while others must not be allowed because they violate basic principles. Sometimes, changes in cultural practices require political power because ethical values have to be made into law. Universal bioethical standards have to be more than promises and principles. Promises and principles have to be made juridical. Ethics at the level of visions has to get down to details and make changes in medical practices which benefit patients in whatever culture they happen to be immersed-. For this to happen professionals need to be taught respect for a universal ethics and need to be supported by professional medical associations with real political clout.

Medicine can provide proof that a universal ethics is possible. It can also set an example for humane treatment of patients across cultural divides. Groups like "Doctors Without Borders" and "Physicians for Human Rights" are already setting a powerful example of how universal standards can be applied across cultural divides. Their doctors remind us of the medical implications of universal human rights. Medical professionals are the "priests" of today's world. If they effectively "preach" a universal ethics by the way they treat patients in every culture, maybe civil society everywhere can be improved. Maybe people locked in Tribal cultures which distort their view of others can be freed. Maybe the human family and a more humane medicine will survive.

VI. Natural Law and Sexuality

EPISTEMOLOGY AND ETHICS

Radical Protestant reformers held the view that salvation is achieved *sola fide*. No matter what the sin, no matter how harmful or hurtful or destructive the behavior, faith alone will wipe clean the soul and guarantee salvation. Good works do not save a person, and evil works only highlight the salvific power of faith. Faith solved all moral problems and lowered concern about morality.

With a somewhat similar mind-set, enlightenment and post-enlightenment secular believers hold to a doctrine of *sola scientia*. No matter what the behaviors of human beings, there is a remedy. Now, however, the remedy is scientific rather than religious. Just as faith overcame immorality in a Protestant tradition, science overcomes immorality in an enlightenment tradition. Science, in this secular belief system, not only has the power to keep things from happening, but when bad things do happen, science has the power to set things right. Science can overcome evil. It can even remedy the harmful consequences of destructive science and technology. As long as science and technology flourish, there is no need for concern about morals and ethics, about immoral and unethical behaviors, about ominous consequences from science and technology. Science does for bad behavior in today's secular world what faith did for sinful behavior in the 16th century Protestant world.

The bottom line is that if we have science, we don't have to worry about morality, and we don't need religion. If there is a problem, science can fix it. Science gave us penicillin for infectious disease.[70] And scientifically informed legislators can devise laws and policies to control hate, crime, addictions, etc. Of course, people without science can have ethical standards, but these are expressions of personal feelings and are similar to poetry, literature, and religion. Guidance for conduct in a post-modern secular culture comes from science.

In a scientific secular perspective, the reality which we confront as human beings is not a created something which poses questions and provides answers which are accessible to poets as well as scientists, philosophers as well as mystics. Rather, reality is just a cluster of quantum particles which bump into one another and produce consequences which over long periods of time result in evolutionary changes. Human behavior or morality requires physics in order to be understood properly. Underneath the enlightenment message of individual autonomy is a set of assumptions which seem to make freedom impossible and presume to explain both human life and human action by hard science. Newton and Locke and Hume provide the background assumptions for our contemporary secular belief system.

In the Catholic tradition, reality was always the foundation of ethics, but reality is understood differently. The right and the good require mind and an understanding of the structure of reality and then respect for that structure. The task of understanding reality is not reducible to bits of matter put together by chance. It is far more complex and never finished. Moral standards are grounded in the very structure of reality which poses questions to human perceivers and makes answers to serious investigation possible. There is a sense in which morality in this tradition is based on physics, but not physics reduced to bits of matter randomly clashing. Rather, the reality on which Catholic morality is based is a created, structured, objective, intelligible reality which can be accessed by the mind of a philosopher, a mystic, a poet, or a scientist.

In a post-enlightenment scientific perspective, sexuality is the movement of particular particles or fluids. It is purely physical and understandable by physics. If we want to understand sexuality, we have to understand the movement of fluids or the different ways in which the fluid movements are expressed in behavior. In other words, we do a physics of sexuality.

A NATURAL LAW ETHICS OF SEXUALITY

In Catholic Natural Law Theory sexuality has a structure and a built-in meaning. Both can be accessed by serious study but not simply by examining fluids. Poets, mystics, philosophers and theologians have all recognized the linkage of sexuality with love and intimacy, with serious commitment and procreation. Human sexual behavior has social, psychological, spiritual, and even religious dimensions. A reasonable sexual ethics takes all aspects of sexuality into consideration.

Reality is not reducible to particles or fluids in a Natural Law perspective, and it cannot be re-structured or re-shaped to fit the desires of a particular culture. Scientists in their tradition have to try to understand reality with their own intellectual instruments and background assumptions, but they cannot reduce reality to random particles and they must respect its intelligible structure. That intelligible structure is the foundation both of truth and of morality. Truth and morality are influenced by personal and cultural perspectives, but they are not the products of particular persons or particular cultures. Truth and morality are grasped rather than created. Neither is meant to conform to the desires of persons living in a certain time and place.

If sexuality is seen in a narrow, reductionistic, post-enlightenment perspective, there is no basis for internationally-recognized moral standards like those which condemn female circumcision, or rape, or sexual abuse of children. And yet there is an instinctual revulsion created by such behaviors. "So what," the reductionist physicists would say! Revulsion, too, is reducible to fluids and movements of fluids. But if a particular culture freely decides against these behaviors, scientifically informed legislators can create effective restraints for them.

Revulsion has a special place and function in the post enlightenment period. Horror films and frightening images play an important moral role in our secular culture. Only horror suggests a limit to behavior when human actions are reducible to particles and fluids. That is why horrible deeds and scenes have such a prominent place today in TV shows and movies. Some critics have expressed concern about horror films and horrible behavior, "but not to worry." "When the culture decides to control or eliminate the horrific behaviors, it can do so with the aid of scientifically-engineered correctives." So goes the post-modern, secular, scientific response to widespread concern about the disintegration of morality in U.S. culture.

In the Natural Law tradition on the other hand, horror and revulsion are a primary and instructional indication that the behavior which causes them is wrong. They are an instructional first step in a moral analysis of a behavior which can then be shown to violate the created structure of reality. Revulsion is more than mere fluids or material particles clashing. It is a natural psychic response to a serious violation of the natural order. Sexual abuse of children causes revulsion in sane and healthy people because it is such a terrible violation of the way children should be treated, based on an objective understanding of what it is to be a child. Female circumcisions and rape are similar violations of women and their bodies. Doctors practicing humane medicine are guided by such natural instincts.

In the Catholic Natural Law vision, there is more to casual sex and rape, violations of women and children, than just material particles and fluids. Treating people, as well as behavior, in purely material, physical, indifferent terms not only leads to horror and revulsion, but to suicides and personal breakdowns. The incidence of suicide among persons who are violated or who violate their own human physiology and psychology and spirituality with destructive behavior is astronomical.[71] This seems to escape many secular people who insist on seeing all reality in the most reductionistic materialistic terms. Only narrow, uncritical fundamentalistic beliefs would explain such blindness. Humane medical practitioners stand for life and so what they can to stop suicide.[72]

CONSERVATIVE SEXUAL MORALITY

If post-enlightenment secularism is reductionistic as far as sexuality is concerned, so too is Natural Law Theory as understood by ultra-conservative and ultra-orthodox Catholics. Natural Law as it is commonly used by such thinkers is too narrowly focused. Oftentimes, in official Church teachings, human nature is collapsed into a narrow physiology. Moral rules are issued based on a respect for nature in the sense of a limited and segregated area of physiology. The advantage of a Natural Law perspective is that it provides an objective foundation for morality and avoids alternatives which claim either that there is no right and wrong or that an action is right and good as long as it is freely chosen. But in order to be convincing, an objective foundation of morality must adequately reflect the complexity of reality and not distort that complexity.

In official Catholic (Vatican) rather than academic moral interpretations of Natural Law, the distinction between *secundum naturam* and *contra naturam* tends to be all too clear. *Secundum naturam* is action or behavior in conformity with nature in the narrowest physiological sense. For example, human intervention into nature is permitted, but only as long as the intervention corrects or improves a procreative function of sex organs. Physiologically viewed, sexual intercourse is reduced to nature's system for reproduction. It is all about eggs and sperm and the system for bringing these two cells to procreativity.

In current official Catholic teachings, sexual intercourse is right and good for married couples as long as the couple does not interfere with reproductive physiology. Only married couples should have sexual relations, because only they could care for the possible offspring. And the only way married couples can ethically control reproduction is to have relations during physiology's infertile periods. Such sexual intercourse is *secundum naturam*. Anything else, like using birth control technologies, is considered *contra naturam*.

The inadequacy of this perspective is evident to most married persons. Families simply cannot manage the complex challenges of maintaining a loving relationship and controlling procreation within the moral parameters of this inadequate view of the nature of sexuality and sexual relations. Inadequacy runs into absurdity when Church officials attempt to respond to tragic situations like the rape of nuns which was common in Africa in the 1960s.

The Vatican at that time decreed that "morally" the nuns had permission to use contraceptives as a defense against rape. The contraceptive, according to this reasoning, could only be used preventively. If rape occurred which the nun had not anticipated, and she was impregnated, then she had to bear the child. She could not use a post-rape contraceptive. This reasoning centered on nature in the sense of narrow procreative physiology. Nuns who could foresee being raped could use contraceptives which inhibit the sperm from fertilizing the ovum, but nothing more. If the rape took place without anticipatory contraceptives, then the nuns had to take the consequences because nothing could be done morally to interfere with the "natural" process.

How unreasonable! How insensitive to the terrible situation of women who were being raped. And the absurdity is "justified" by Natural Law arguments which reduce the reality of sexuality to sperm and ova, while ignoring the broader reality of sexual relations. The one official exception to the requirement that all sexual relations remain open to procreation (nuns permitted to use contraceptives when they anticipated being raped) shows just how unrealistic and irrational this narrow ethical perspective is. There is no mention of forced sex being *contra naturam* and the need to offset its consequences. The only consideration was a narrow sexual physiology and the only moral concern aws that relations be open to procreation. Some conservatives cannot even believe that the Vatican made an exception to its "Natural-law based rules." Ordinary married Catholics see the exception as one more glaring proof of the inadequacy of the moral teaching and the unreasonableness of its theoretical foundation.

The official Catholic view of nature is too narrow, and consequently, the moral teachings based on it are counter-intuitive. Physiology has to be taken into account when trying to understand a Natural Law-based sexual morality. Procreation is an important aspect of sexuality and has to be taken into account. But a Natural Law perspective is not just about the physiology of sex organs and procreation. Nature, and especially human nature, is the objective foundation of moral teachings and human beings are persons. There are physical, psychological, social and spiritual dimensions of human reality, including human sexuality. All have to be taken into account when trying to decide what is *secundum naturam* and what is *contra naturam*. A narrow view of nature which undermines or ignores the broad role of sexuality in human life and loving human relationships is not a defensible basis for a Natural Law ethics of sexuality.

One need only look at official Church documents which address issues of sexuality to see this narrow and distorted perspective and to see how little of the reality of sexuality is reflected in the official moral teachings.[73] Human sexuality is depreciated by being reduced to and dominated by its reproductive dimension. Official moral teaching about sexuality has little to say to lay persons because it is so little concerned with ordinary sexual experience and indeed with ordinary life. Pope John Paul II and conservative moralists who follow his lead ignore the insight of Thomas Aquinas in his *Commentary on the Nicomachaen Ethics*. "What pertains to moral science is known mostly through experience."[74]

Some conservative Catholics think that John Paul II's teachings on marriage and sexuality are prophetic. If prophetic means being against today's secular culture, then the Pope is prophetic. Like some prophets of old, he consistently uses a style of condemnation and asserts his truths rather than arguing them. In that sense, he is definitely more a prophet than a moral theologian. But if prophetic means addressing ordinary people about the issues of their lives in language that they can understand and with arguments that are convincing, then John Paul II is more like other mainstream conservatives. He talks in a way that few understand and has little or nothing to say that will help ordinary married couples better understand and appreciate their sexuality.

Official Catholic moral teachings tend to reduce the complexity of body experience to sexuality, and then to reduce sexuality either to sinfulness or acceptability depending on whether sex remains open to procreation. If one were to do a survey of the millions of former Catholics who now are Protestants, the vast majority would refer at least to the narrowness of official moral teachings on issues of sexuality. If celibate men are the main producers of official Catholic moral teachings, then it is altogether understandable that the teachings will be unrelated to the experience of married people. If sexuality in secular culture is not something serious and very permissive, in official Catholic cultures there is too much seriousness and almost everything sexual is sinful.

A DIFFERENT CATHOLIC SEXUAL ETHIC

Many Catholic ethicists would like to reform official teachings. At least they would like to provide a respectful alternative, one based on nature in a much broader sense. An alternative Natural Law ethics of sexuality is based not on the reproductive dimension of sexual physiology alone, but following St. Thomas Aquinas, on the broad and complex sexual experience of married lovers. For married couples, sexual relations can be charming and casual, clumsy and awkward, serious and silly, the expression of affection and a way of making up after a fight. Relations can be lustful or restrained, exploitative or an act of generosity. Centering moral teachings on reproduction is like looking at marital sexuality from outer space. The reality of sexuality in ordinary experience has a wealth and richness which is absent from official moral teachings. Catholic ethicists have to look at human experience in order to understand sexuality and in order to construct moral teachings on sex. Couples need the positive contributions which sexuality can provide when reproduction is not wanted or needed or responsible or even possible. In an alternative perspective, there is more to the objective structure of human sexuality than reproduction, and being open to procreation is not the base condition for moral sex.

In a liberal Natural Law perspective what a particular mind can grasp about body and sexuality is never the whole picture. Serious thinking and sound investigative methods can provide some answers to questions raised by sex. But the answers will never be complete. Even sound moral reflection cannot answer all the questions. Human beings never understand everything, and this is true especially of persons who try to write about difficult issues. Catholic moral teaching about human sexuality has to be modest. It presumes to be neither unchangeable nor infallible. It remains ever open to improvement and perhaps correction as our understanding of nature broadens.

A modest and evolutionary understanding of sexuality in a liberal Natural Law perspective takes sexual experience seriously. Sexual experience poses questions but also provides answers and better understanding. Celibates who focus on corporal and sexual control do not see the whole picture because they lack experience. We know from the moral teachings of Catholic conservatives that control is necessary. Sex in contemporary culture does seem to be out of control. And yet we know from non-official Catholic understanding that control isn't everything. Today's sexual horror films (e.g., *Silence of the Lambs* and *Hannibal*) also make a case for control. But where is the case being made for letting the experience of married people instruct us about the structure of sexuality and about its moral limits? Certainly not in the ethics which informs official teachings.

If experience tells anything about sexuality, it speaks of pleasure. But where is the natural reality of pleasure given its due recognition in official and conservative moral teachings? One would get the impression that this intense and obvious dimension of sexuality were a distortion or a sin. Does this make any sense? Does it ground ethics upon reality and experience? Can it qualify as Natural Law thinking? Couples who stay married don't do so by restricting and rejecting sexual pleasure but by indulging and fostering it. Successful sex and multiple orgasms are as contributory to successful marriage as anything. Why not celebrate sexual pleasure in Catholic moral teaching? Why all the emphasis on control?

Marital sex in the official Catholic teachings is dominated by fertility cycles. The all-important positive benefits of pleasurable sex are all too often repressed by concern about fertile periods and by the dangers of fertility. Fertility is important, but fertility isn't everything. What are unmarried or separated or divorced people supposed to do about their isolation? What about homosexual people? What about elderly widows and widowers who are especially pained and depressed by the loss of bodily contact and with another person? Shouldn't a Natural Law-based ethics of sexuality address these issues? Official teaching would make one think that there were no such problems, or if they arise, then the gritting of teeth is the only moral response. Sexuality, however, is more than a means of transmitting life. The other dimensions of sexuality need to find expression in a Catholic moral system based on objective reality. A broader and more open perspective on sexuality might even welcome some sexual imagery in Catholic liturgy.

The broader, more open perspective, however, cannot abandon the challenge of drawing lines and establishing limits for sexual behavior. Sexuality, for example, must support and protect the family, without which human beings cannot grow physically, mentally, emotionally, morally. Sexual behavior which undermines, weakens or threatens the family context would be *contra naturam* and wrong. Incest is an obvious example of unethical and unnatural behavior. It is wrong not just because it destroys the family environment which is crucial for human development. It is wrong, too, because it does harm to most vulnerable persons and violates individual dignity instead of fostering love and care. Incest is the very opposite of caring and nurturing behavior and therefore violates the nature of sex and sexual morality. The same is true of pedophilia, and yet this reality was missed by bishops in their handling of the priest scandals. These behaviors violate the objective nature and structure of sexuality. Arguments based on the consequences of such behavior are strong, and they support an objectively-based Natural Law ethics of sexuality.

Instead of a moral perspective which focuses on each individual sexual act, what about a perspective which focuses on attitudes and dispositions? What about a virtue or character ethics instead of an act ethics? Can anyone make any sense of the claim that each individual sexual act using a contraceptive technology is *intrinsically* evil? Even killing another person may be an acceptable act under some circumstances; e.g., in order to save one's own life. Can anyone believe that using a contraceptive in all circumstances is intrinsically evil? Would the use of contraceptives destroy the intimate and unitive aspects of marital sexuality? If not, how could birth control technology be intrinsically evil?

Those who teach the official morality spend their time trying to teach young married couples how to walk around their reproductive biological processes. The focus of their teaching is on reproductive physiology, and the immense richness of sexuality is either ignored or repressed. What about all the other uses and meanings of sexuality? Are intimacy and unity served by giving up sexual relations during fertile periods when they are most pleasurable and most beneficial to the relationship? What about the importance of relieving women's worries about becoming pregnant? How about sex for celebration instead of procreation as a moral purpose for marital relations?

A more reasonable Catholic perspective on sexuality would support the use of birth control by couples trying responsibly to control procreation. (See chapter on birth control.) It would strongly support condom use to prevent the transmission of HIV. (See chapter on AIDS.) Their type of Catholic thinking would follow St. Thomas Aquinas and rely on the experience of married couples for understanding sexuality and the morality of sexual relations. It would also focus more attention on women's experience. We have heard enough about how celibate men understand sexual morality. It's time we hear from good laypersons actively involved in sex. It's time for Natural Law morality to get in touch with reality. It's time for the Church's moral teachings to address the people's experiences, to help them with the issues which trouble them, and to develop moral teachings which are reasonable and convincing.

Catholic Natural Law bioethics starts with experience and works toward understanding the structure of sexuality and then toward the construction of a sexual morality. Catholic ethics is objectively based. It does not turn its back on what others have thought and argued in history, but it continually reviews even long-standing moral teachings in light of new experience and new insights. It is neither radical nor revolutionary but unrelentingly critical. And it does not hesitate to speak out when official teachings no longer make sense. Liberal Catholic bioethics has to establish limits and to say what is *contra naturam*, but it does not speak too quickly. Reasonable ethics requires patience as well as courage. It has to be willing to learn even from laypersons in the Church who have been ignored as a source of insight. Especially in the area of sexuality and sexual ethics, the Catholic perspective has to be open to change. But how does one draw the line between *secundum naturam* and *contra naturam* in this broader and more liberal understanding of human nature? That is what the conservative perspective does best. (This challenge is taken up in the chapters on technological limits.)

Post-enlightenment scientism and conservative Catholic teachings are rejected by most people today because both seem too confining. Both fail to take into account the experiences of intelligent people. Post-modern culture was embraced by some because it offered an attractive alternative. But this alternative easily leads to rampant nihilism. The official Church's insistence that people bow to its teaching authority while giving few reasonable arguments to support the teachings played into the post-modern argument that religion represents unjustifiable authoritarianism and philosophical obscurantism.

Human Beings seek meaning and reasonable standards in their lives. While many secular people abandoned post-modernism, many Catholics turned their backs on the Church because the moral teachings on sexuality seemed cold and unrealistic. If the Church were to broaden its approach to Natural Law, taking into account new knowledge and marital experience, it could be in the position of offering meaningful answers to life's problems. It would also be positioned to profit from a reaction against post-modern emptiness

Openness to change in moral teachings is not openness to any novelty but rather openness to the changes going on in every culture at the beginning of the new millenium. Catholics have left the Church in droves because of the irrelevance and irrationality of its moral teachings. And sexual morality is not the only issue contributing to empty pews and increasingly aging congregations. It is not just Catholicism which is losing members. In a *New York Times* book review, one of the writers estimated that during the 1990s as many as 50 churches a week shut their doors. Vacant churches are everywhere in evidence in the American countryside. In the cities, closed churches are converted into everything from nightclubs to supermarkets.

It would be an exaggeration to blame an increasing American secularism on unreasonable moral teachings about sexuality alone, but certainly these are an important cause of this phenomenon. What official Church documents have to say about sex and family is further and further removed from the reality in which most people live. The old ethnic Catholic neighborhoods and parishes are either gone or on the way out, and with them the values and older understandings of marriage and sexuality. Not all these changes are healthy, and the official Church rightfully takes stands against certain sexual practices which are pervasive in today's culture. But a Church totally against the culture is more like a sect. It gives up on the more difficult task of staying in dialogue with cultural changes and practices. It gives up trying to provide direction for the broader culture and moral teachings based on nature that the mass majority of people can understand. If the Church were a business, what is happening today is equivalent to failing to protect its market and its market share.

People who remain Catholic still have questions about morality and sexuality, but they no longer go to the Church for answers because what the

Church has to say is either unconvincing or unreasonable. Church officials seem to move from one self-inflicted wounding to another without ever seriously looking at the continuing stream of sexual scandals within the Church and the awful consequences of official handling of the scandals.

Church authorities are understandably more comfortable repeating the traditional stands on sexuality and family, but at some point Church teachings have to enter real dialogue with what is going on in 21st century culture, the messy reality in which modern people live. Scientism and post modernism and Catholic fundamentalism are all failures. They cannot provide moral direction. Catholic fundamentalism may provide a comfortable retreat for a while from all the moral messiness, but it has no respectable intellectual foundations and cannot meet people's needs. The Church has to take some stands against the culture, but it cannot be totally anti-culture and remain Catholic. A connected and relevant Catholic Church has to support its moral teachings with convincing reasons. Authoritarianism won't do. It will only increase the membership flight. If not this alternative Catholic perspective, at least some components of a more liberal Catholic vision will either be part of Church interaction with the world, or the stream of dissatisfaction with the Church will become broader and deeper.

NATURAL LAW SEXUALITY AND HUMANE MEDICINE

If post-modernism's reduction of sexuality to physics works neither as an explanation nor as a source of moral direction, the same is true of conservative Catholic moral teachings on sexuality. In conservative Catholic moral teachings it is the reduction of sexuality to procreation which neither explains the reality nor provides moral direction. In today's world, believers and unbelievers alike search for understanding and moral direction in this area of life. Both secular and religious parents have to provide moral direction regarding sexuality for the next generation. The contemporary situation provides an opportunity for churches to step in and to fill a role which religion occupied for centuries: i.e. the source of moral direction. In fact, however, religion tends to be a last place people look for moral direction regarding sexuality. Strange as it may seem, moral direction for Catholics, at least at an academic level, has come from medicine. Humane physicians have provided both convincing understanding and effective moral direction in the area of sexuality.

Americans do not ordinarily look to physicians for moral direction. Historically, however, doctors were moralists and for sound philosophical reasons. The foundation of morality in classical philosophy was nature (physis), and the expert on the structure of human nature was the physician. Because the good was grounded on the real, and medicine studied the human reality, physicians provided both medical therapy and moral direction. Doctors guided patients on how to live their lives and exercised moral authority over them.

Since doctors knew the body and its structure, they were qualified to provide moral direction grounded on that human nature.

In contemporary life things have changed. Today ethics and bioethics especially in the U.S. are differently grounded. Science, pleasure, consequence, autonomy are some of the more popular modern foundations of morality. In Catholic culture however, ethics continues to be grounded in the rationally accessible structure of reality. Natural Law ethics continues to derive the good from the real. It is human nature or the human reality that provides the foundation for what is determined to be good. Good and right contribute to human fulfillment. If the issue is sexuality, then the nature of sexuality is the ground for what is good and ethical. It is therefore not all that amazing that distinguished Catholic physicians, obstetricians, and gynecologists would step forward to provide direction in the form of sound rational arguments for change in moral teachings based on an updated and more adequate understanding of nature and the structure of sexual reality.

Dr. John Rock

Dr. John Rock was born in Marlborough, MA, and in the 1920's he started a fertility clinic. Throughout his professional life he was concerned with the needs of couples who were childless and his research concentrated on the first stages of conception and on disorders of human reproductive physiology. Dr. Rock's experimentation over many years led to the discovery of the oral contraceptive pill.

Dr. Rock was a physician devoted to his church and devoted to his patients. He was both a good Catholic and a humane physician. He was a scientist who understood the anatomy and physiology of sexuality and he had, as well, a rare competency in moral philosophy. He not only discovered the birth control pill but also defended its use by well-articulated argumentation based on Natural Law theory.[75]

Dr. Rock knew about sexuality not only from personal experience with a beloved wife but from dealing all his life with women's concerns with aspects of sexuality, especially with the possibilities of pregnancy and childbirth. He looked carefully at all the complex anatomical and physiological factors in sexuality, but looked carefully as well at emotional and psychological dimensions expressed in the personal needs of the women whom he treated. He was a humane physician and it was his humaneness joined to his medical competency which compelled him to work to discover an effective way of aiding the unreliable natural system of birth control. He came up with an effective birth control technology and sound explanations based on Natural Law theory for its moral legitimacy. I didn't know him personally, but from what I have read about him and his work, I believe he was a practitioner of humane medicine and a saintly liberal Catholic.

Dr. Rock made convincing arguments for the satisfaction of legitimate sexual needs and for the use of sexuality beyond procreative purpose. He distinguished between enduring abstract principals of the Natural Law and concrete applications of the principals in changing situations which justify evolution in moral teachings. Practical concrete moral judgments, he argued, have to benefit for accumulated insight and have to be open to change.

Reasonable and convincing moral direction in the area of sexuality came from this physician who understood the structure of sexuality. From his medical understanding joined with his personal experience and that of his thousands of patients, he provided a solid foundation for some needed changes in the Church's sexual ethics. Sad to say, his reasonable, practical Natural Law-based moral direction in sexuality was ignored by church authorities.

Dr. Andre Hellegers

Historians of the discipline of bioethics all recognize the important influence of another liberal thinking and humane physician: Dr. Andre Hellegers. In the first bioethics period, post World War II, the ethical issues with which philosophers and theologians were preoccupied had to do with death and life. The most pressing life issues were abortion and birth control. The birth control issue was pushed into public scrutiny by the Supreme Court decision in *Griswold vs. Connecticut* in 1965, and by increasing pressures on married couples to control family size. Discussion about the ethics of contraception forced into consideration the nature of sexuality and the related issues were addressed by Protestant, Catholic, Jewish, and Secular Ethicists.

No Catholic ethicist addressed the ethical issues surrounding sexuality more passionately and intelligently than Andre Hellegers. During the sixties, he published more than 50 articles in learned journals and in the print media. Many of the articles were about birth control and world population. He was one of the few laypersons invited to participate in the Papal Commission on Population and Birth Control, called to reconsider the church's moral teachings about these issues. When Paul VI overturned conclusions of the experts, Dr. Hellegers wrote a critique of the Encyclical in *Medical Opinion and Review* (5:71, 1969) He followed up with other writings, including a "Scientist's Analysis," in *Contraception, Authority, and Dissent* (New York: Herder and Herder, 1969) Dr. Helleger brought his concerns about sound moral direction regarding sexuality, especially regarding birth control to meetings with respected theologians and publishers. Out of these discussions emerged one of the founding institutes of bioethics, *The Kennedy Institute of Ethics* at Georgetown University. The issue of birth control had a lot to do with his concern about science and morality and his reasons for working to found the first university based bioethics institute. In the words of Warren Reich, Andre Hellegers was intensely invested in linking scientific progress and unfettered moral debate.

Before the Enlightenment, religion was the principal source of morality and moral directives were imposed from above, either by a monarchial government or by a religious monarchy (the Pope). After the Enlightenment, in secular society, pluralism replaced homogeneity in moral matters. Individual autonomy gradually became the basic moral standard (right or good is what I choose to decide). In Europe, science, especially the social sciences (economics and sociology), attempted to fill the gap left by the collapse of monarchies and the separation of church and state. Marxism and Capitalism both offered a new moral vision for society.

Religion did not disappear as a source of moral teaching and direction but it could no longer be imposed authoritatively. Even moral teachings provided by church authorities had to make sense to the faithful in order to function as real directives. The important Protestant moralists in the 20th Century had moved to accept a change in birth control teachings, but the change was much more difficult for the Catholic hierarchy who still presumed to exercise control over Catholic moral teachings. Under Pope John Paul XXIII liberal Catholic thinkers argued for approval of contraceptive technologies which would make birth control more reliable. Andre Hellegers was a prominent member of this group.

When the Papal commission called by John Paul XXIII failed to move the next Pope (Paul VI) to change the traditional teaching, Andre Hellegers saw the need for developing convincing and supportive reasons for moral directives not just for contraceptive technologies but for or against the explosion of medical technologies generated by government investment in medicine after World War II. In 1971, Dr. Hellegers succeeded in attracting economic support from the Kennedy Family and then attracted some of the most talented philosophers and theologians to establish the *Kennedy Institute of Ethics* at Georgetown University. His institute was the first university based center for the study of what by then was referred to as bioethics. The Kennedy Institute scholars addressed the expanding issues generated by scientific medicine. The new discipline of Bioethics quickly moved beyond issues of sexuality but the need for rationally defensible moral direction in sexual matters certainly influenced the beginnings of the discipline. The original name of the *Kennedy Institute of Ethics* was *Joseph and Rose Kennedy Institute for the Study of Human Reproduction and Bioethics*.

In the early period of the Kennedy Institute, Andre Hellegers the humane physician joined with Albert Jonsen a distinguished theologian and together they published an essay in which they reached out for what could be described as a modernized Natural Law foundation of ethics.[76] Later he supported and encouraged publication of an Encyclopedia of Bioethics, an enormously important work carried out by Warren Thomas Reich, one of the scholars with whom he invited to establish the Kennedy Institute.

In the absence of credible moral direction coming from official religious teachings, it was the discipline of medicine and liberal theologians who stepped in to provide moral direction. Shortly after the establishment of the new discipline of bioethics, female scholars joined the field and kept attention focused on the issues of interest to women including sexuality, birth control, and pregnancy. (Loretta Kopleman, Karen Lebacqz, Renee Fox, Ruth Macklin)

LIFE

VII. HIV/ AIDS and Condoms

At a confidential meeting some years ago called by WHO, I consulted with a large number of Catholic bishops from Africa and Latin America who were gathered to explore ways in which the Church could help with the AIDS crisis. In some of the African dioceses, HIV infects as many as 25 percent of the population, and AIDS threatens to devastate both Church and society. The very continuation of life in some dioceses is threatened because young people are afraid to marry and start a family for fear that they or their partner might carry the disease. In these communities, some of the poorest in the world, even testing for AIDS has been too expensive for national health-care budgets. The American drugs for slowing AIDS development are economically out of reach. A large percentage of males are migrant workers and spend considerable time away from home. In such circumstances, protection against the spread of AIDS becomes extraordinarily difficult.

The bishops clearly wanted to do something. They knew that the use of condoms has proved to be the most affordable means of containing the spread of AIDS. But given the Vatican's public stands on sexual ethics, they knew that their options were very limited. They asked whether the threat of AIDS creates circumstances in which the use of condoms in marital intercourse may sometimes be justified even within the context of Catholic moral teachings about sexuality.

TRADITIONAL CATHOLIC MORAL THEORY

To address these questions requires, first, a quick review of the elements of Catholic ethical theory, then an application of the theory to the case of condom use, with special attention to the approach consistently taken by Vatican spokesmen. The argument may hold little fascination for people not affected by Church authority, but the stakes involved in a successful resolution of the questions are extremely high for African people. The worst pandemic in history remains out of control. The Church's credibility and intellectual integrity are at risk. So are human lives, in great numbers.

Mainstream Catholic ethics, or more properly, the Natural Law tradition in ethics, in its essentials is followed by many thinkers who are not Catholic. Natural Law Theory is rooted in classical Greek philosophy and shares with it an approach to morality that is dependent upon a concept of *physis* or *natura*. Nature for the Greeks was something objective and living which served as the measure for determining right and wrong. *Right* meant harmony with nature; wrong was a violation of nature or lack of harmony. Nature was considered the seat of divinity as well as morality; to go against nature violated both religion and ethics.

The interpretation of *physis* in matters having to do with the body and its functions was the domain of Greek physicians, who were also moralists. They prescribed with authority over every aspect of their patients' bodily lives: food, drink, rest, work, sexual relations, bodily excretions, exercise, sleep, living conditions and even emotional and psychic states. They did so because all these elements (*res non naturales*) could cause illness, which the physician was obliged to prevent and sometimes cure. Physicians' directives were right when they restored nature to proper functioning or prevented illness in the sense of bodily imbalance. Basic medicine was preventative medicine. Therapeutics aimed at restoring the balance of nature's proper functioning. Only when preventative and rebalancing measures failed did the Greek physician prescribe a *pharmacon* or drug, and only when drugs failed was cutting or surgery considered morally right. Historically Nature, in the sense of the body's natural balance and proper function, was the standard of moral right and wrong.

In both Greek philosophy and Catholic ethics, and more recently among some environmentalists, this standard applies as well outside of medicine. Technology, for example, is judged right and good accordingly as it promotes nature, brings nature to fulfillment, respects nature's functioning, or imitates nature. It is bad when it violates or destroys the natural order. What is morally right is based not solely on whether the intention of the actor is good or her behavior is free, but on something more objective, a natural order that exists "out there" and can be studied to discern what may or may not be done.

Acceptance of this traditional Natural Law approach does not preclude disagreement about its application to situations today. The classic concepts of *physis* and *natura* lend themselves to widely different understandings, so that historically, Natural Law Theory has rallied forces for both revolutionary change and reactionary intransigence. (See chapter on Natural Law, History and Politics.) Broadly speaking, one school of thought within the tradition (which may be labeled "conservative" or "traditionalist") tends to defend the *status quo*; its proponents identify natura with their memory of the way things had been, and they tend to limit the application of Natural Law to issues of individual morality. Those called "liberals or progressives" are more likely to use Natural Law to set standards for the way things ought to be in the future according to a natural order more broadly interpreted (a bio-psycho-social reality). Understanding nature to include concepts of a rightly-ordered human community, liberals tend to focus more on issues of social justice than on issues of personal morality.

INTERPRETING CATHOLIC LAW

Where issues of sexual morality such as birth control, homosexuality, and the use of condoms are concerned, conservatives usually focus on what they understand to be the natural structure and function of individual physiological parts. Less conservative Catholics prefer to extend nature to include its higher functions. They focus on the "nature of the person" or "personhood" as the objective standard of morality and include socio-political considerations for understanding the human reality. A morally-right act, for example, would respect the autonomy and uniqueness of persons, or it might bring personhood to fulfillment in a certain socio-political context.

What holds these differing interpretations together is a conviction that nature itself, and especially human nature, serves as the ground for discerning what is right or wrong. The Catholic tradition stands over and against an ethics of intention, pure subjectivity, autonomy, an ethics of good will, an ethics of consequences, a post-modern relativistic ethics.

Although both conservative and progressive Catholic moralists use some version of St. Thomas Aquinas' standards for making a moral determination, (the *act*, the *intention* and the *circumstances*), there are differences of emphasis. Conservatives prefer to focus on the "the act," downplaying both intention and circumstance. They fear that giving weight to intention will lead to subjectivism and that consideration of circumstances can foster relativism. Liberals, on the other hand, contend that the act itself can be properly described and morally evaluated only by giving careful attention to intention and personal circumstance as well as history and culture and social circumstances. Differences of intention, for example sometimes help to determine both the nature of an act and its moral quality.

All these differences of interpretation and emphasis were very much on display in the battle over the morality of birth control that went public with Pope John XXIII's creation in 1963 of a high-level commission to study the issue. A strong majority of the members of the commission (later enlarged by Paul VI) eventually recommended approval of certain contraceptive technologies, viewed now as respectful of nature or as AIDS to a natural procreative process, more broadly understood. In the face of an accumulation of expert opinion, however, Pope Paul VI reaffirmed the traditional teaching in his 1968 encyclical *Humanae Vitae*, arguing that only marital acts which physiologically were "left open" to the transmission of life were natural and morally-right acts. Decades after the encyclical, the issue continues to split the conservative from liberal theologians and for that matter, to split the great majority of married Catholics from Church authorities. In the context of the AIDS epidemic, critics ask whether marital acts in some African contexts must be left open to the transmission of death. (See chapter on birth control)

In the birth control debate, it was largely the discovery of "The Pill" by Dr. Rock that brought a reconsideration of the morality of birth control. Now it is the existence of AIDS, a disease for which there is as yet neither a cure nor a vaccine and which so far appears to be invariably fatal, that urges moralists to consider the relevance of Natural Law to the use of condoms for disease prevention.

MORALLY EVALUATING CONDOM USE

The question of condom use holds little urgency for those uninformed people who still think of AIDS as a disease confined to "gays." But in Africa and most of the world today, AIDS is a heterosexual disease, and transmission is primarily through heterosexual relations. In the U.S. and the rest of the developed world, the incidence of HIV infection among heterosexuals is measured in the millions, and it will continue to rise exponentially. As awareness of this reality spreads, sympathy for AIDS victims may rise, along with fears of personal vulnerability.

Some years ago, the news media reported the death of Ryan White, a young hemophiliac who contracted AIDS through a blood transfusion and was subjected to meanness and ignominy at school because of his condition. Ryan's suffering helped many people see AIDS as the human tragedy it is and not as conservatives still think of it as a divine punishment. Whatever the source of an AIDS victim's infection, however, (through transfusions, drug abuse, accidental exposure of health-care workers, or sexual encounters), the medical evidence makes clear that condoms, though they do not give complete protection against sexually transmitted infection, nevertheless contain disease transmission and offer the best chance for saving millions of lives.

For liberal Catholic ethicists who dissent from the teaching of *Humane Vitae*, the fact that condoms may serve a contraceptive end does not require that their use be proscribed under every circumstance, e.g., when the decision of married couples to space or prevent procreation is taken for good reasons. The problem is different for more conservative theologians, however, and for bishops. Can conservative Catholic moralists admit that the use of a condom for disease prevention is different from condom use for contraception? The condom is the same, but its function is very different, as different as procreation is from disease transmission. Used as contraceptives, condoms work by blocking sperm from making contact with the ovum. Used as a disease prevention device, the condom works by blocking the deadly virus from making contact with another's bodily fluids. The HIV virus that leads to AIDS is carried not by the sperm, but by the white blood cells or lymphocytes within the seminal fluid, or in less concentration, in the vaginal fluid. A disease barrier is not a procreation barrier even though procreation is indirectly affected.

To analyze these issues in a hypothetical context, imagine the case of a hemophiliac man who may become infected with HIV virus through a needed blood transfusion. Both he and his wife need sexual intimacy to maintain their marriage and family. They have a child and would like to have another. They are not anti-life nor do they harbor a "contraceptive mentality." But the husband understands his obligation not to infect another person, and the wife knows her right to protect her own life. Hemophiliacs never know whether a transfusion will contain a virus which escaped through a blood test window of vulnerability. For them, condom use is the only way to prevent infection of the wife while nourishing their love and satisfying their sexual needs. Needless to say, this hypothetical situation is duplicated many times over in real life. Thousands of health-care workers are exposed to HIV infection by accidental sticks and splashes, and they need to protect their spouses until they are sure about their HIV status. Not to expose another person to harm is the right thing, even if one's own exposure came about through use of a nonsterile needle in drug abuse, or through sexual contact with a male or female prostitute.

One approach to the moral question is to bring into play the classical principle of double effect. Applied to certain situations over the years, this principle has provided uncontested justification for needed medical procedures that would not otherwise be allowed. For persons infected with the AIDS virus, the act of protecting a partner from deadly harm is a right act which has two effects. As the example just cited shows, it is easy to imagine circumstances in which the contraceptive effect is recognized but regretted. In some sexual relations, the condom use itself as well as overt, direct, and exclusive intention is to protect another person from death or to preserve one's own health while maintaining family unity through marital intimacy during a difficult time. Achieving those ends through condom use prevents conception as well as infection, but so do certain medical procedures long accepted as entirely licit. Removing an infected uterus, for example, may bring about permanent infertility; but unless that result is directly intended, the procedure is acceptable under Natural Law Theory as it is understood by both conservative and liberal interpreters.

In the case of AIDS, there is no procedure, medical or surgical, which can cure the disease, only a procedure to prevent its spread. But preventative medicine is as valid as therapeutic medicine; in fact, it is generally more respectful of the body's natural mechanisms. Condom use, too, removes or inhibits transmission of an infectious agent; in some circumstances, it enhances and strengthens physis and natura in the sense of supporting nature's own preventive strategies, when ordinarily protective immunological mechanisms are too weak to create consistent barriers against infection.

Circumstance (including consequence) is all-important for the description and definition of acts and their moral evaluation. Using a condom to prevent the spread of infection may superficially appear to be the same act as using a condom for contraception. Obviously, the condom is the same. But pulling the trigger of a deadly weapon in order to commit murder and pulling the same trigger of the same gun in self-defense against a murderous assailant, though physically identical and employing the same device, are not the same. One fragment of a complex human behavior does not define the whole behavior. Murder involves much more than killing. Murder is always wrong and in some cultures punishable by death; killing in self-defense, however tragic, is morally justifiable and not punishable at all. Once it is recognized that circumstances enter into the definition and the moral evaluation of actions, then for persons with every right to have sexual relations, condom use is not merely justifiable but obligatory.

Conception is not the same as infection, and protecting against one is different from protecting against the other. The device used in both cases is the same, just as the weapon may be the same in acts of self-defense and murder. But the act is different: different elements of the behavior constitute its definition. If circumstances or contexts alter act description and definition, and they do, then moral evaluation also changes. This argument concludes that, in some circumstances, condom use qualifies as a right, sensitive, respectful, and nonmaleficient act. Condom use in fact can rightfully be called pro-life.

Even conservative Catholic moralists who continue to consider condom use to be evil, regardless of intention and circumstances, must take into account the moral obligation to prefer a lesser evil when the choice is between two evils. For centuries, moral theology tracts have provided examples of choosing the lesser of two evils. If someone has already decided to take another person's life, it was always considered proper (for example) to encourage killing his cow or setting his barn afire instead. If an infected person is going to go on having otherwise licit sexual relations, then the lesser evil would be protective sex. Even if unmarried persons have decided to have sexual relations and one is infected, it would be doubly wrong not to use a condom. A lesser hurt, a lesser damage, a lesser violation has always been considered morally acceptable when the alternative is much worse, and surely the use of a condom is less wrong than infecting another person with a lethal disease.

BISHOPS ON CONDOMS

Such arguments, however cogent they may appear to most lay Catholics, have not won acceptance from many in the Catholic hierarchy.[77] Consider the relatively moderate statement, "The Many Faces of AIDS: A Gospel Response," issued in 1987 by the administrative board of the United States Catholic Conference, and some of the reactions it engendered. Its section on preventing the spread of AIDS contained the following moderate argument:

> "Because we live in a pluralistic society, we acknowledge that some will not agree with our understanding of human sexuality. We recognize that public educational programs addressed to a wide audience will reflect the fact that some people will not act as they can and should; that they will not refrain from the type of sexual or drug-abuse behavior that can transmit AIDS. In such a situation, educational efforts, if grounded in the broader moral vision outlined above, could include accurate information about prophylactic devices or other practices proposed by some medical experts as potential means of preventing AIDS. We are not promoting the use of prophylactics, but merely providing information that is part of the factual picture. Such a factual presentation should indicate that abstinence outside of marriage and fidelity

> within marriage are the only morally correct and medically sure ways to prevent the spread of AIDS. So-called safe sex practices are at best only partially effective. They do not take into account either the real values that are at stake or the fundamental good of the human person."

It seems conceivable that the episcopal authors of the statement were influenced by their acceptance of pluralism, reflected in the abandonment many years ago of Church efforts to keep on the books state laws forbidding the sale of contraceptives. In any case, the statement in no way departed from traditional official teaching on sexual morality. It said nothing, for example, about the situation of a faithful husband or wife accidentally exposed to and infected by the HIV virus. From its silence it would appear to follow that Catholic couples in such circumstances are required to choose between perpetual celibacy and a high risk of fatal infection. And yet, referring to the document, Archbishop (now Cardinal) Anthony Bevilacqua said that "some have even suggested that Catholic moral teaching on the use of contraceptives may no longer be as firm as it was in the past." Other bishops, also in response to the statement of the administrative board, issued their own statement condemning condoms absolutely. In their words, "it is never morally permissible to employ an intrinsically evil means to achieve a good purpose."

This last statement is at best a serious abuse of language. The phrase "intrinsically evil" is properly used to describe only the most repugnant acts, so repugnant that they could never be justified, no matter what the circumstances, or no matter what beneficial consequences might result from performing them. (It is interesting that the phrase is most often used in matters of sexuality: not for nuclear war, not for willful large-scale destruction of the environment, not for the death penalty.) Some Catholic bishops are so careless as to appear to describe a loving and love-affirming sexual act between a husband and wife in the circumstances described here as intrinsically evil; that is, to place it in the same category with acts that are unimaginably and irredeemably evil. This is not defensible, in any assessment faithful to the Natural Law tradition which is grounded on a rational analysis of behavior *within a context*. It is possible only for ideologues and their too-faithful disciples. And there is, in fact, an ideology at work here, one held by some few theologians and many bishops who occupy influential positions at the Church. This ideology, so narrowly focused on sex and sexuality, is not only bad Natural Law thinking, but suggests the presence of a psychic pathology.

A biblical fundamentalist does not consider the context of a scriptural passage (its literary mode, purpose, context, relationship to other passages, or to the whole work in which it appears), and consequently may adopt interpretations that are both dogmatic and irrational. If a biblical text says that "God is a rock," the fundamentalist, if pressed for interpretation, may insist on a literal meaning: "If it says, 'God is a rock,' who are you to say otherwise?" The Catholic tradition, both in biblical exegesis and moral philosophy, requires consideration of context and rejects mindless, unreflecting, literal interpretation. The notion that a condom is a condom and no context would ever justify its use reflects the same mindset which insists that God is a rock, and that a rock is a rock, a mindset apparently seriously ill and far removed from the Catholic tradition. And yet, just such irrationality has become identified with the Catholic Church in many people's minds because of uncareful episcopal pronouncements on sexual matters. Such stands serve to undermine the reasonableness of religion at a time when the Church must compete with a host of secular meaning systems. In the case of condom use, pathological teachings also inhibit important efforts to control the spread of AIDS. One must look to Church politics and to the grip of radically conservative views in some quarters to explain this scandal.

UNHEALTHY MORAL VIEWPOINTS

Many examples of extremism in high places have been cited over the years in newspapers and journals. In Chile some years ago I read a long (three full New York Times-style pages in small print) article in a Chilean newspaper, *El Mercurio*, which profiled an important Church official. Bishop Carlo Caffarra is a close consultant of Pope John Paul II and widely considered to be his "point man" on matters of sexual morality and the family. At the time, he headed an institute founded by the Pope at the Lateran University dedicated to research on family matters. His mindset and moral convictions on a range of issues came through clearly in the long personal interview. Bishop Caffarra saw the Church as involved in a death struggle with the modern world; for him it is a struggle between good and evil. Again and again, he referred to the modern technological world as the "corrupt empire," a phrase reminiscent of the rhetoric of George W. Bush, except that Caffarra wraps both Marxist and Capitalist societies in the same diabolical cloak. Contemporary European culture he referred to as "the great lie." Catholic priests and theologians who disagreed with his view he called, "priests and theologians of the corrupt empire."

The engine driving the corrupt empire, in Caffarra's view, is technology. "The laws of the machine, which is the law of matter," he believes, "is being applied by man to himself and it will reduce him to slavery." It follows that he levels an absolute condemnation against any technological means of family limitation. Birth control, reproductive technologies, and abortion are lumped together; all are symbolic of the evil modern culture. To speak of using contraception to avoid abortion makes no sense, for the two are inseparable: "Contraception is the first step toward abortion," he said. People who use the "natural" birth control method, he believes, will keep a child that happens to be conceived, but if contraceptive technologies are used, an unexpected pregnancy is terminated by abortion.

Caffarra's views on AIDS were predictable. Like all other technologies which "interfere with nature," condoms are evil; they "belong to the culture aligned with death and make death man's friend instead of his principal enemy." Diffusing information about the use of condoms to prevent HIV infection not only increases homosexual relations, but actually serves to spread AIDS. For Caffarra, this is a truth that is never mentioned in the media, and "woe to him who mentions it." All this accords with the prism of categories, (black and white, good and evil, God or the devil), through which he sees the world. Faust, Mephistopheles, the Antichrist, Wagner, Marxism, modernity, utilitarianism, and nihilism: Caffarra resorts to these names and concepts repeatedly in his commentary on sexuality, the family, and the modern world.

This is an astonishing mindset, and one that is frightening in a man who influences Catholic moral teachings worldwide. Similar views, expressing a siege mentality in approaching the modern world, can be found in pronouncements from other Church leaders, though couched in less apocalyptic terms. Democratic values and respect for diversity are viewed with suspicion by conservative Church officials; and when they are seen to be permeating the Church, frightened conservatives direct their energies not to dialogue with faithful Catholics who cannot in conscience accept their radically negative evaluations, but to simple repression. Connected with irrational moral perspectives about life and sexuality are totalitarian political structures and styles in today's Church.

As already noted, the consequences that may follow from applying an ultra-conservative mindset to the issue of condoms for prevention of AIDS infection are anti-life and simply unacceptable. The unreasonableness of the prohibition of condom use is evident enough in the case of the hemophiliac marriage or the accidentally infected health-care worker. To say that all marital acts must be open to the "transmission of life" is to require that they must also be open to the "transmission of death." To hold that the only acceptable alternative for millions of persons is sexual abstinence, condemns people already under sentence of death to reject not only the pleasures of sex but the consolation of intimacy at a time of ultimate stress and to require the same of his or her partner in marriage.

CONCLUSION

In societies where the AIDS virus has created a plague, it becomes yet more irrational to view the use of condoms as "intrinsically" evil without regard to intention, circumstance, and the character of the act. Does the Church, which rightfully prides itself as standing on the side of life, stand by and observe the dissolution of life in parts of Africa and Asia? One way to continue human life in threatened areas is to permit couples in the first period of their marriage to make use of the condom until they can be certain that they are free of infection, and only then to start their families. The condom in such situations is not an antiprocreative device but the precise opposite. Used as a temporary protection against possible infection, it serves as preventative medicine, as well as the only plausible and practical way to help assure that children, when they do come, remain uninfected.

This alternative perspective on Catholic moral theory, setting forth as it does the possibility of justifying condom use in certain situations, supports what common sense insists must be done in many tragic circumstances around the world. If a vaccine or a cure for AIDS is found and made available globally, there will be plenty of time for further academic analyses of sexual morality in professional journals. For now, the challenge is to control the spread of AIDS. The use of condoms for that purpose is fully justified within the Catholic Natural Law tradition.

POSTSCRIPT

In April 2000, a Vatican theologian (Monsignor Jacques Suandeau) published an article in the Vatican Newspaper, *L'Observatore Romano*, in which he started by admitting that the official Vatican teachings on condoms and AIDS had been widely criticized for lacking sense of reality and for being irresponsible. The criticized Vatican teaching was that condoms are intrinsically evil and should never be used even if the purpose was to protect against transmission of HIV and other sexual diseases.[78] The Monsignor made a distinction between prevention and containment which permitted him to deny that condoms alone will prevent HIV/AIDS. He could, however, admit that condoms have had success in containing the disease in certain populations. He also admitted that condoms are a "lesser evil." This term he used to justify or permit what formerly had been proscribed. The article did not admit a mistake in official Catholic moral teachings. But it did show that change even for conservative Catholic moralists is possible.

VIII. Papal Authority and Birth Control

Before the term bioethics was introduced, there were issues which preoccupied ethicists, physicians, and the whole culture. They were issues related to the beginning of life, and the birth control issue was one of these. Primitive birth control techniques have been in use forever. Once Charles Goodyear launched the rubber industry, more sophisticated technologies appeared. By the mid-19th century, there were condoms and diaphragms and cervical caps and IUD's. Later, pharmacies offered chemical contraceptives in the form of suppositories, vaginal sponges, and medicated tampons.

Moral objections to all these devices were based on the separation of sex and procreation. By 1873, certain moralists were powerful enough to lobby for a congressional bill which described contraceptives of all sorts as obscene and prohibited their distribution both by mail and across state lines. Twenty-four states passed similar legislation. Manufacturers of birth control technologies were legally vulnerable, but courts frequently offset prosecution based on privacy.

In 1918, a New York appeals court ruled that contraceptives were legal because they were useful for disease prevention. Following that decision, new manufacturers appeared who produced millions of birth control devices a day. In 1965 the Supreme Court, in Griswold v. Connecticut sanctioned contraceptive use based on privacy. Since the 1960s, Dr. Rock's birth control pill has been in widespread use. The issue of birth control in the 20th century became a critically important one for both married couples and for moral theologians.

This chapter was written in the mid 1960s, (during Paul VI's reign) and must be read with an awareness of this timeline. It is included to show that the arguments on one side and the other have not changed much over the years. And, it is included to make the point that in the issue of birth control had everything to do with the development of bioethics as a separate discipline.

THE STATE OF CHURCH TEACHING TODAY

The Catholic Church has for some time been in a state of doubt about the question of birth control which, euphemistically, is being referred to as a moment of study and reflection.

The whole thing happened without warning. After years of enforced silence and uniformity on this issue, the Church erupted with dissent from every side. Philosophers began a broadside and convincing critique of the Natural Law base of the traditional teaching. The laity started expressing their disagreement in impressive numbers. Most unexpectedly from the floor of the Vatican Council came voices of high-ranking ecclesiastics making impassioned pleas for change in the Church's teaching. In the midst of cries for change from one side and absolute denials of such a possibility from the other, Pope John XXIII called a special commission of theologians and laymen to consider the matter. News stories about this commission did no more than substantiate the differences of opinion which everyone knows exist within the Church. Those for and against change developed their reasoning. Reading the different opinions demonstrates how widely divergent opinion was and still is on this issue. The different opinions provide testimony to the fact that the old certainty and uniformity about this moral teaching no longer exists.

Uncertainty is not always bad. It can and has been the impetus behind a flurry of studies and discussions which have substantially improved our understanding of this issue. Respected theologians have posed the problem differently and come up with answers that are radically different from the traditional ones. Uncertainty has led to the development of a reasonable and convincing alternative to the Church's traditional teaching. This in turn is putting tremendous pressure on the officials of the Church to recognize the fact that there are now two very different moral opinions. Such a recognition by itself amounts to a big change and should be reason enough to abandon the claim of required assent to the official teaching.

Pressures for change are coming as much from what Catholic people are doing as from what theologians are thinking. Life must go on even while the Church authorities struggle with change. While the official Church cogitates, decisions are being made and actions are being taken. Seventy percent of today's Catholics think that the Church should change its stand on birth control, and 65% of married people under 35 are practicing some form of birth control.[79] There is really no need to refer to professional polls to substantiate these statistics. The untrained eye is besieged by evidence.

The old Catholic family of eight, 10, and 12 is as rare as the horse and buggy. The two, three, and four member family is everywhere in evidence, and it takes no genius to make the proper deductions. Increasingly large numbers of married couples are taking the matter into their own hands, and they are supported by theologians who insist that a doubtful law does not bind. Despite official proclamation to the contrary, the doubt is obvious. Forming one's own conscience in such a situation is considered fully justified, indeed, the only reasonable response.

The way good Christians form their consciences is important in determining the morality of this issue. The Holy Spirit does not confine activity to the higher levels of ecclesiastical hierarchy. Every Christian is the recipient of the Spirit. Married couples, prayerfully searching out the proper way to express their mutual love and fulfill their Christian vocation, definitely come under the influence of the Holy Spirit. The way good Christian people behave plays just as important a role in developing moral teachings as the way Christians in the past interpreted the revealed message and contributed to dogmatic teachings. The Church is more than a priestly caste. Moral teachings, especially, involve the entire Christian community. Unilateral decisions on highly complex problems made in isolation from the community are not only not infallible. They are not even reasonable. They violate the foundation of Natural Law ethics and stand an excellent chance of being wrong.

CATHOLIC ETHICS CHANGES

A moral act is a free human response to a particular situation. Judging an act to be right or wrong requires that the situation be addressed, and this may amount to a bewildering complexity of factors. Value claims grow out of such assessments, and the values themselves change when the factors upon which they are based change. The traditional rules against birth control, for example, reflect a set of values flowing from a situation that is no longer with us. When the world was under-populated and economically organized around agriculture, human reproduction merited high value. There was more than enough room for everyone. A new life in that situation was a blessing, especially when the family supported itself from the land. A high infancy mortality rate made a high reproduction rate a necessity for species survival and thereby further enhanced the value of procreation. In Old Testament times, the greatest blessing was a large family and the worst curse was sterility. The value of procreation was reflected in the prayers, hopes and ceremonies of the Hebrew people.

The situation today has greatly changed. We live neither in an under-populated world nor in an agricultural society. Can we continue to talk of population increase as a value, when in fact it has become a menace? In Old Testament times, it took 1600 years to double the population of the world. Today it doubles every 35 years. There are too many people to feed and an entirely new economic system. Rather than an economic asset, each child represents an enormous economic responsibility that cannot easily be met. Most families must be squeezed into city dwellings and provisions must be made to amass the small fortune required to provide a child with the minimum mental and physical equipment needed to function in the world.

The new situation or the new reality is an all-important factor in making a moral judgment about birth control based on Natural Law. To continue to insist on moral values and rules formed in response to a set of circumstances which no longer exist cannot be justified in Natural Law thinking. To elevate a moral value relative to a certain age or situation to the status of an absolute is unreasonable. The population explosion, the pressing responsibilities of the modern family, and recent developments in moral theology must all be considered when we set out to make a Natural Law-based judgment about birth control. Today, we human beings are in an entirely new game, and this calls for a new set of rules.

All talk of changing situations or of an evolution in morality is considered beside the point by defenders of the traditional Catholic teaching on birth control. No circumstances can change the morality of an act considered to be intrinsically evil. This moral judgment, they claim, is based on "unchanging principles of the Natural Law." There is an enigma here. The one characteristic of a judgment based on the principles of the Natural Law is that it be clearand evident to decent persons with normal intelligence. The Natural Law's objection to efficient and effective birth control, however, is evident to an ever-diminishing minority of ultra-conservative Catholics alone. This enigma warrants a closer look at the concept of Natural Law and its application to the question of effective birth control.

THE ROLE OF REASON IN CATHOLIC MORALITY

Natural Law Theory has had its ups and downs in history. Surprisingly, it has survived as a useful way of viewing both individual and social moral problems. Its endurance is a testimony to its substance and depth, but it has not endured unchanged. The concept of Natural Law has undergone continued development and refinement, especially under the influence of contemporary insights about the meaning of human existence, the place of love in marriage, and the totally different situation caused by an overpopulation which strains the very survival of God's creation.

The Natural Law Theory stands for a morality based upon something more objective than individual feelings and broader than ethnic-based values. Besides the attitudes and restrictions which individuals choose to impose upon themselves and the moral rules legislated by society, there are certain principles of conduct which bind all human beings. Natural Law is concerned with principles of moral conduct accessible to every person who enjoys the use of reason and normal moral maturity.

The general principles of the Natural Law are derived from what reason shows us about human nature and the requirements for developing, respecting, and sustaining it. Basic human nature manifests itself in the fundamental tendencies and drives which we find wherever we find human beings. Reason indicates that good for humans exists in the proper fulfillment of these basic drives and tendencies. Natural Law, then, is the rational ordering of one's acts toward human fulfillment and away from what undermines moral, intellectual, emotional, social and religious development.

This law is called "natural" because its principles are arrived at by considering the character, make-up and nature of human beings and the universe which humans confront. General principles are formulated in relation to goals toward which the given patterns of human nature and basic human tendencies are directed. Considering that human beings are reasoning beings, one general principle would be: "Be rational" or "Use your reason according to its proper structure." In its negative form, the same principle would proscribe unreasonable acts: i.e., acts which are destructive either of human life or the natural order. Following from the recognition of the uniqueness of each individual human being would be the general principle requiring respect for this uniqueness expressed in any number of ways. Immanuel Kant stated this principle in terms of never using a human person as a means to something else. Sartre would say, "Never make a human being into an object." Buber would insist on human relationships being "I-thou," and not all "I-it" forms.

Reasonableness applies to human acts toward other humans and toward the world. Love is one of the basic needs of human beings, and marriage celebrates this need. Marriage corresponds to human nature and requires that marital relations be carried out according to a developing understanding of all that love and marriage involves. Actions which destroy love or frustrate marriage or prevent it from attaining its diverse ends (among them procreation) would clearly be considered wrong by Natural Law.

The general principles of Natural Law enjoy a broad transcultural recognition because most everyone trying to understand human life and moral behavior agrees with them. It is difficult to conceive of any situation in the future in which the principle requiring humans to be rational and to love and to respect the uniqueness of others would be invalidated. The perennial character of these moral imperatives is based on that which remains the same in human existence throughout development and despite cultural differences. The same is not true, however, for the application of principles to the intricate and complex practical situations in which humans must make most of their moral judgments. As one descends from the general principles to the concrete situations, from the abstract world of clear and distinct ideas to the muddled and knotty ambiguity of practical life, the certainty and universality of what reason requires disappears. Discredit has been visited upon the whole concept of Natural Law because some of its advocates try to make every judgment of reason as evident, unchangeable, and certain as the formalistic reasonable principle. The truth is that neither Saint Thomas nor any other reputable Natural Law theorist claims any such clarity or certainty for judgments of reason in concrete, complex, ever changing situations.

Regarding marriage, reason shows us not only that this institution corresponds to human nature, but that it is necessary for human survival. Mutual fulfillment and procreation, which are important dimensions of a marital relationship, must be respected. One can be firm and secure in insisting upon respect for these dimensions. One cannot, however, be equally firm, secure, and certain about every question or every behavior of the married couple. If marriage is naturally ordered to procreation and continuation of the species, then it must be permanent and stable because human offspring need dual parenting. But the fact that marriage is ordered or structural to procreation does not mean that it is ordered only to procreation or that every marital/sexual act must be so ordered. Particular questions of marital morality can be so cluttered with factors which must be considered, assessed and balanced against one another that what is reasonable must always remain open to revision.

The more concrete the situation, the more elements are involved that reason must assess in order to make a right judgment. Such complexity militates against perfect clarity or self-evidence. Saint Thomas in his exposition of the Natural Law insists that constant investigation and re-thinking are necessary to arrive at what it requires. The more concrete the issue, the farther reason's dictates are removed from certainty and universality.

> "Since the speculative reason is busied chiefly with necessary things which cannot be otherwise than they are, its proper conclusions like the universal principles contain the truth without fail. The practical reason, on the other hand, is busied with contingent matters about which human actions are concerned; and consequently, although there is necessity in general principles, the more we descend to matters of detail, the more frequently we encounter defects".
> (*Prima Secundae*, Question 94, Article 4).

Saint Thomas held that not even monogamy could be shown to belong to the unchangeable and absolute requirements of the Natural Law. In certain situations, he held, exceptions could be made. Monogamy, however, seems much more obviously required by reason for the fulfillment of nature's goals in marriage than biological openness of each sexual act to procreation. If particular circumstances require different moral judgments regarding monogamy, then it would seem that St. Thomas, were he around today, would have no trouble allowing for change or at least differences regarding traditional teachings about birth control, especially in light of the almost totally new situation in which modern people find themselves.

Exceptions even to general principles are justified in Natural Law Theory. If the principle of self preservation which obliges us to respect our own biological integrity can be set aside for higher reason, (giving one's life for the good of another) there is every reason to expect that the principle requiring respect for the procreative dimension of marriage and marital sexuality might also admit of exceptions. Defenders of the traditional Church teaching about birth control make an absolute of something that is open to change in light of changing circumstances. Some conservative Catholics insist that procreativity in sexuality be respected in splendid isolation from all other principles and valid no matter what the circumstances. This is unreasonable. In a particular case, there may be any number of extenuating circumstances, as well as a coalescence of many other principles obligating a person to economic, social, psychological and spiritual responsibility. Human beings are more than their procreative biology, and a Natural Law Theory requires that we take into consideration the many other dimensions of human beings.

Natural Law requires a reasoned response to human being and the reality which humans confront. Natural Law reasoning requires not just an assessment of a situation, but an assessment made in light of the values and goals which are enshrined in human being and in the created order. Principles and circumstances interact in the formation of particular moral judgments. This interaction, however, can never be such that the Natural Law judgment is no more than a cold, mechanical application of a principle to a situation.

The act of moral judgment in a particular situation is both unique and creative. The prudential act which judges what to do in a particular situation must make up for what is unclear with a unique personal initiative. That to which all human beings are obliged morally can never be outlined in such detail that all the peculiarities of each concrete case are covered. There is always room for the personal prudential factors and individual discretion. The individual judgment must always be present in moral choices, and its sphere must be guarded against those who would equate conscience or morality with a logical conclusion from stated premises or a passive obedience to some law. Natural Law ethics leaves room for prudence and freedom of individual judgments about particular circumstances. It leaves room for conscience.

The concept of Natural Law has been distorted beyond recognition by those who would resolve the most complex, practical questions about how to achieve responsible parenthood by a snap reference to a traditional teaching and then insist that the proscription of artificial birth control in a particular case is certain, infallible and irreformable. Nothing but confusion has resulted today from those in the Church who claim that the traditional teaching on birth control cannot change because it is based on Natural Law. They would apply the universal, certain and unchangeable characteristics belonging only to the most general principles of Natural Law to judgments about particular matters that are far removed from general principles and far removed from the certainty, universality and unchangeableness enjoyed by principles.

Natural Law Theory as developed by moral theologians from St. Thomas and even before St. Thomas permits no such exaggeration and extrapolation. The case for or against technological birth control therefore is not closed. It is by its very nature an open question. It is not only open because human nature is open, free, self realizing and developing but because the many factors involved in concrete moral judgments about such complex issues as procreation and love making can always be seen in a different light and assessed in a different way. No one judgment in a matter of this sort can be the last word. This is so no matter who makes it.

Contemporary theology has contributed its share to the pressure for change in traditional Church teaching precisely by shedding new light on the value of marital love and sex. Marriage has traditionally been looked upon as a contract for the exchange of acts conducive to procreation. Today, marriage is much more reasonably considered as a humanizing, personal relationship built around the love of a married couple. The love of each partner reveals the other as he or she really is and welds the two together in a unique bond. Marriage is a love union characterized by fidelity. It means, "I love you and you alone, forever." It is a paradigmatic "I-thou" relationship.

Within marriage understood in these terms, sexuality becomes an all-important expression of this self-creating love. The place of sexual love in marriage can no longer be talked of in terms of legal rights and duties any more than marriage itself can adequately be described in legalistic, contractual categories. Sexual acts in marriage are the natural expressions and guarantees of a union which Saint Paul uses as an analogy to exemplify Christ's loving union with His Church. Sexual acts are not the mere fulfillment of a legal obligation. They continue to have meaning and importance even when the procreation possibility is either already fulfilled or no longer possible. The all-important expression of love belongs as much to the meaning of sexuality as does reproduction, and the old moral rules which emphasize only the physiological integrity of the act are bad examples of Natural Law reasoning.

BIRTH CONTROL TECHNOLOGY AND CHURCH AUTHORITY

As traditional moral teachings about the nature of marriage have been challenged, Church officials have tended more and more to argue their case in terms of Church Authority. Their reasoning most often runs along the line that the Church has spoken and therefore the matter is settled.[80] This is an unusual twist for moral judgments supposedly based on rationality and what reasonableness requires. The Church definitely has spoken on this matter, not once, but many times. In every instance, however, the Church statements were based not on divine revelation but on reason, and therefore presumably were accompanied by convincing, rational argumentation.

Arguments from scripture shed little light on the nature of marital relations and no light on the problem of artificial birth control faced by contemporary married couples. The subject is not even mentioned in the New Testament. The Torah records the story of Onan in Genesis 38: 8-10, but scriptural exegetes insist that Onan's sin and punishment are related to the Levirate Law which required that a man marry his brother's widow and bring forth offspring in his brother's name. Once the Levirate marriage was accepted, offspring had to be provided for the deceased brother. Onan refused this obligation after contracting marriage with his brother's widow. His punishment is related to this obligation and not to any contraception act. There is no mention of contraception in the Torah despite many minute proscriptions on sexual matters.

Even without direction from Scripture, theologians over the centuries were forced to concern themselves with the issue of sexuality and procreation. According to one theological opinion, sexual intercourse was lawful only when its purpose was procreative. Saint Augustine gave this view considerable weight because of his prestige. Consequently, it was changed only with great difficulty and only after centuries. Logically, this teaching would have made sexual relations with an infertile wife sinful.

Actually, Augustine's view was a liberal response to Manichean theology which divided reality into radically opposite good and evil realms. For them, all matter and flesh were in the realm of darkness and evil. To be moral one had to avoid both sex and procreation. Augustine, in response to this Manichean absurdity approved of sexuality and sexual relations, but only if they were procreative.

Gradually, over the years, Catholic moral teaching was further modified by a distinction between intentional and unintentional infertility. Little by little, reasons other than procreation were admitted as sufficient for justifying sexual relations in marriage (avoidance of adultery, restoration of physical or mental health, etc.). Only by way of this slow evolutionary process was the Church able to overcome the Manichean influences on St. Augustine's moral theology and St. Augustine's enormous personal influence. Because of Augustine's prestige, Catholic moral teachings on sexuality in general and in particular on procreativity have been marked by an unhealthy Puritanism. It took until the nineteenth century for love's expression and fostering to be recognized as moral purposes for marital intercourse in Church teachings. But in every instance, these other justifying reasons for sex had to be present in such a way as not to intentionally exclude procreation.

Development in the Church's moral teachings continued over the centuries and is reflected today in Vatican II's statement on marriage in which conjugal love finally reaches a substantial status independent of procreation.

> This love is singularly expressed and perfected by the proper work of marriage. The acts then by which the spouses intimately and chastely unite with each other are decent and worthy, and exercised in a truly human way (they) signify and foster a mutual giving by which with joyful and grateful spirit they reciprocally enrich one another.
> (Section 49, Scheme 13, *On the Church in the Modern World*).

The Council viewed marital intercourse not in the old terms of sexual organs and the procreative finality of sexual intercourse, but rather as a totally human act contributing to the enrichment of the love between spouses which the Lord requires.

Catholic moral teaching does change. Development and evolution of Natural Law teaching is obviously necessary and is reflected in Church history. Vatican II's statement is a far cry from either Saint Augustine or the medieval scholastics with their teleology of organs and classification of primary and secondary ends of marriage. This broader more personalistic view gives to marriage for the expression of love a new independence from the procreation fixation. The criterion for judging the morality of marital sex has correspondingly been greatly broadened. Addressing married couples who have to decide about sexual acts, theologians at Vatican II said the following:

They will thoughtfully take into account both their own welfare and that of their children, those already born and those which may be foreseen. From this accounting they reckon with both the material and spiritual conditions of the times as well as their state in life. Finally they will consult the interest of the family community, of temporal society and of the Church herself (*Church in the Modern World*, II, c.1 par. 50).

Council documents, however, did not make a total break with the Augustinian tradition. There was an appendix attached (probably by dissenting conservatives) which explicitly stated that "Every deliberate intervention of man which vitiates the work proper to the conjugal act of the person is contrary to the divine law and the order of matrimony." Later Paul VI in an allocution to his cardinals referred specifically to birth control saying, "we must say openly that up to now we have no sufficient reason to consider the norms of Pius XII in this matter to be out of date and therefore not binding." They continue to bind, "at least until we feel in conscience obliged to modify them because in a matter of such importance it seems good that Catholics wish to observe a single law." (The norm of Pius XII forbade artificial birth control, and we see how difficult it was for Paul VI, an admirer of Pius XII, to change. Any such change would be hard to justify in the Vatican political culture.)

In the 1960s Pope Paul VI added another word on this issue in his encyclical on Poverty (*Populorum Progressio*). This progressive document recognized the tie between economic development and the control of population. "The size of the population increases more rapidly than the available resources and things are found to have reached an apparent impasse." The Pope recognized population control as a justifiable area of government concern as well as the right of the government to intervene "by favoring the availability of appropriate information and by adopting suitable measures, provided that these are in conformity with the moral law and that they respect the rightful freedom of married couples." There was no specific mention of any change on what constituted Church law on birth control, but there was a strong and pointed reference to the role of the consciences of married partners in determining the number of children they could reasonably support. Conspicuously absent from this paragraph was any reference to "artificial methods," or "unnatural practices," which had so often been mentioned in Papal statements. Here again development is obvious, but on the heel of the encyclical came a warning that this development was not to be viewed as a change in the Church's stand on artificial contraception. After the publication of the secret reports of the Papal commission, another warning was issued in the *Observatore Romano* (the Vatican newspaper) against presuming that there had been any change in the church's official teachings.

Those who stand to be the most shocked by any change in the Church's moral teachings are Catholic conservatives. Liberal Catholics would welcome change and find it easier to respect Church teaching authority. If change does not occur despite overwhelming reasons for it, then most Catholics hardly consider themselves outside the Church for disagreeing with traditional teaching. They understand and recognize the need, indeed the inevitability of evolution in moral teachings. For them dissent within the Church is also good and necessary.

After the Council spoke so eloquently about freedom of conscience, it would be absurd to think that this freedom applied to everyone except Catholics. The exercise of the sacred right of conscience cannot disqualify one from membership in the Catholic community. The unity of the assembly must certainly be founded on certain truths held in common by the members, but this does not require absolute uniformity on every issue. There have always been differences within the Church on dogmatic as well as moral questions. To make conformity on birth control necessary for membership in the Church would be equivalent to putting birth control on the same level as the Incarnation and the Trinity.

LIBERAL ATTITUDES TOWARD CHURCH TEACHINGS

Besides having sympathy for dissent within the Church, liberal Catholics will tend to look at any official Vatican statement in historical perspective. The Church has changed time and again on matters of morality, and in light of this development no single stand can be considered a final one. The liberal may even consider a particular papal pronouncement to be a regrettable mistake, but this will not be a valid reason for leaving the Church. The mistake may very well have been caused by a bad formulation of the question, and liberals might work toward a reformulation of the question. The advancement of contemporary philosophy and theology toward a better understanding of human life, marriage, and sex cannot help but contribute to the correction of what can reasonably be considered a skewed Manichean influence on Catholic teachings regarding sexual morality.

If the Pope, however, were to make a statement reaffirming the need for responsible parenthood and insisting that the married couples themselves have the responsibility for forming their consciences as to the proper means, such would require a difficult adjustment for traditionalists. They might very well consider such a development in the magisterium to be in direct and irreconcilable opposition to former statements on this matter. Tending to consider every papal statement infallible or almost so, such a stand would be equivalent to an infallible statement to the effect that former pronouncements had been fallible. This would be a difficult corner to escape from.

Even if embarrassed by a papal pronouncement, Catholic liberals could cite many examples of statements arrived at on the basis of a bad formulation of the question and then subsequently corrected or re-interpreted by later pronouncements. The formulae of Pius IX in his famous *Syllabus of Errors* are a case in point. Under the pressure of bad times for the Vatican state and the disadvantage of a suspicious and negative attitude toward the emerging enlightenment concepts of freedom of conscience and religious liberty, Pius IX condemned the proposition that "…it has been quite right to provide by law in certain Catholic countries that foreigners taking up residence there should be allowed to exercise their own particular forms of public worship." (No. 78) This is a far cry from the statement of John XXIII and the council's declaration on religious liberty. Later teachings specifically required the very things which Pius IX considered it necessary to condemn. There have been sufficient pronouncements between these two statements for anyone to recognize development in papal teachings.

There is, however, no denying that the end result of evolution and development in Church moral teachings may be as close as one can get to a complete about-face. From the above-mentioned example we can draw two important conclusions: (1) Every papal document must be looked at in its particular historical setting. Cultural and historical pressures in an age might cause an over-reaction on the part of a Pope that later will require correction. Historical and cultural factors may just as well be responsible for a faulty formulation of the question that can obfuscate rather than clarify the issue discussed. Another word, then, is always to be expected. (2) Obviously each papal pronouncement cannot be considered infallible. Theologians are hard put to come to an agreement on any one papal pronouncement that falls into this category. Certainly a papal statement made today on the complex question of birth control would not be considered infallible by anyone, the Pope included.

A unilateral statement by a Pope on sexuality might be full of scientific insight and historical sense, but it will of necessity lack the important element of experience.[81] Experience is reality talking back to us, and we drastically need this dialogue to decide issues of morality. If there is one single bad consequence from use of a Natural Law perspective, it is the tendency to produce a rationalism (deducing moral solutions from abstract principles). If there is any one counter-balance needed to correct this rationalistic tendency, it is a strong dose of empiricism (the development of moral norms starting with an accurate analysis of what people are doing and how it affects them) mixed with historicism (seeing each issue in its historical context). Catholic sexual morality has been all too evidently the work of celibate clerics, and its greatest weakness shows up where the morality makers lack experience. It is just not reasonable to expect priests or bishops or popes to make a series of cerebral deductions from principles and come up with well-grounded moral teachings about sexuality for persons involved in complex marital situations. Only experience in marriage and with marital sexuality can give the needed balance to Natural Law teachings in this area.

Certainly there are instances in which experience is unnecessary for coming to a well-rounded and functional understanding of a problem. A doctor need not actually experience a disease to be able understand its structure, recognize its presence and successfully prescribe for it. Is there not a parallel here to what the Church authorities have done regarding marriage and marital sex? The answer, I believe, is no.

Marital love and lovemaking are not diseases. A disease can be studied from the outside as a question to which an answer can be found. Love can never be understood from the outside. It must be experienced to be fully understood. Surely there is some knowledge we can come to from a purely objective study of marital love, but the results of the study will always be abstract and perhaps unrealistic to the one who actually has the experience. The understanding of marital love and its expression in marital sex from the outside is like understanding suffering from the outside. It lacks an all-important dimension. Not having been experienced, it is never truly known.

The prescribing of concrete moral regulations for the proper expression of marital love by celibate clerics is comparable to the drafting of rules for the monastic life by laymen who have never spent time in a monastery. Experience in many cases is not something that can be dispensed with. Where love, suffering, hope, etc., are the questions under consideration, experience is not just helpful. It is altogether revelatory and altogether necessary. There is truth about sexuality and marital love which is available only to married persons, and when it is absent in lawmakers or moralists, the laws or regulations are bound to be deficient.

MORAL CRISIS AND ECUMENICAL OPPORTUNITY

There is a crisis today in the Catholic Church precisely because moral directives have been imposed which might appear convincing on the philosophical plane but cannot be made to square with the experience of married persons. Married people, forced for one reason or another into the practice of birth control, have found effective birth control in no way harmful to their relationship. Rather, it has been found to be a great boon to their love, their families and their peace of soul. These people cannot be convinced that contraceptive acts violate the nature or meaning of the sexual act, for there has been nothing in their experience to justify such a judgment. Since they are not bad, corrupt, vice-ridden people, but good, sincere, practicing Catholics who live for their spouses and their families, this experience must be counted in assessing the morality of birth control. To refuse experience of this type is to expose moral teachings to error, and what is even worse, unreasonableness and irrelevancy.

Moral regulations that cannot be a model for real life contribute to the erosion of the whole religious structure with which they are associated. Too many good Catholic people, unable to accept the traditional teaching on birth control, are thrown into all kinds of doubts and anxieties about their faith. Too many good people are being forced into a religious no-man's land by irrelevant rationalistic moral teachings joined to an official ecclesiastical intransigence.

No one knows with certainty how this controversy is going to end. We don't even know what the next move will be. Ominous results are predicted for whatever course is followed, and one is hard pressed to be joyful or optimistic for the institutional Church; however an important unforeseen ecumenical benefit may come from this crisis.

The Christian community worldwide has long suffered from the effects of denominational division and interfamily prejudice. Christians badly need to overcome the splits that today are no more than relics of passé theological controversies and long gone political interests. Christianity badly needs to recover its unity, and there must be a place in a reunited Church for the role of Peter, which is too explicit in scripture to be ignored even by biblical fundamentalists who consider church history irrelevant. The papacy, however, as it is presently conceived and structured can never meet current ecumenical needs.

The Pope and papal functions have been and continue to be the victim of a myth. In the popular mind, the Pope is thought of in terms of a sacral figure who enjoys a direct pipeline to God and as a result of this advantage, he issues streams of infallible statements on every conceivable problem. All too many Catholics and even more Protestants think of the papacy in this thoroughly untheological and unscriptural way. Catholics think of the Pope in these terms and are overawed. Protestants think of him in the same way and are horrified.

It would be presumptuous to predict exactly what the functions of Peter's office will be in a reunited Christianity, but certainly, it will be different from what it is at present. The Pope will have to be less ruler and more moderator; less supreme authority and more symbol of unity. The development of this new papacy will take time, perhaps too much time for the urgent needs of today's Church, but the present Pope's handling[82] of the birth control issue could very well contribute to this development. He has already contributed to demythologing the office by his hesitation, doubt, reversals and procrastinations. He appears more and more the man, Giovanni Baptista Montini, shy, retiring, cautious, cerebral, less and less Supreme Pontifex and the heir of the Roman Emperor. All this is for the good. It is altogether possible that the Papal myth, so overburdened with an infallibility syndrome, will finally be laid to rest only by some obviously fallible solemn pronouncement. This could very well come on the birth control issue.[83]

POSTSCRIPT

Following the publication of this article, I was fired from my teaching position in a Catholic seminary and suspended from the priesthood. The seminary where I taught was a vibrant institution with an enrollment of about 200 students.

Following my dismissal, the seminary was closed. None of the empathy patience shown to pedophile priests was shown in this incident. I had challenged authority rather than confess a sin and suffered the consequences. Institutional practice reflects institutional structures and culture. Today, in 2010, there is still no movement toward dialogue with advocates of a more liberal perspective on birth control.

IX: Modifying Abortion Policies: The Role of Metaphor in Medicine and Law

INTRODUCTION - THE UBIQUITOUS METAPHOR

In the touching, tender, and widely acclaimed movie, *Il Postino*, a shy unlettered man, working part-time for the post office, strikes up a friendship with the internationally acclaimed poet, Pablo Neruda. The postman is curious about Neruda's world and the reasons for his fame. He buys a book of Neruda's poems and tries to understand what for him is a new and strange type of discourse. "What is it that makes something a poem?" he asks. "Metaphor," answers the poet. Then, he gives the postman a definition of this term and examples of metaphors even in ordinary speech.

Metaphor uses one kind of idea or object in place of another to suggest a likeness between them: e.g., "a *marble* face." Metaphor is the use of a word or phrase which ordinarily has one meaning to denote something similar but different: e.g., "ships *plowing* the sea." Metaphor is an implied comparison. Metaphor transfers the meaning of one word to another. "Plowing" is something done to soil, but is now applied to water. "Marble" is a type of stone, but is now applied to flesh. In metaphor different realities are compared with one another in order better to understand or to appreciate their complexity.

Understanding metaphor for the postman meant liberation from a flat, superficially understood reality and a barely articulate style of communication. The concept of metaphor gave the postman insight. He started to use metaphors and then moved on to writing poetry himself. Gradually, his whole life changed. He began to understand more, found a way of communicating with a beautiful woman, developed a loving relationship, married, conceived a son, and moved from part-time postal worker to political activist. Once ridiculed as a human failure, he gained fame and widespread respect. A life was changed by understanding metaphor. Even the postman's untimely death turned out to be a metaphor for the tragic dimension of life. Metaphor accompanied both his coming to be as a person and his passing away.

We are used to metaphor in poetic discourse to express a poet's insights. Poets use a word or phrase or image to denote something other than its common or literal meaning. Metaphor, however, is not confined to poetry. Because it is so close to the core of human existence, we find metaphor everywhere we find human understanding. Metaphor is used even in science to flesh out, organize, and communicate raw data which by itself would make little sense, e.g., a black hole, and the big bang. Ordinary life and commonplace discourse are full of metaphor, "a horse laugh," to "hot dog," to "find a smoking gun." Academic disciplines and every type of academic discourse employ either overt or hidden metaphors. In this chapter, I want to introduce metaphor to our speech about ethics in order to enrich the discourse about human nature and human actions.

Applying the general principals and background vision of Natural Law to concrete situations often involve the use of metaphor.

METAPHOR IN LAW

Like poetry, philosophy, science, ethics and law all make use of metaphors in order to understand and then to articulate their visions and judgments. Sometimes for example metaphors are used at trials to convince jurors of a claim or to help them to understand raw facts. An image is created which pulls disparate factual elements together into a meaningful scenario. But these are not the only uses of metaphors in law. Metaphors are absent neither from the formalized statements of law, nor from the academic literature about law. Usually, however, they do not appear explicitly in a legal text, and they remain unrecognized by most readers of legal writing. Even lawyers frequently miss critical background metaphors. They approach legal texts much the same as fundamentalists approach the Bible; focused only on the literal meaning of a phrase or a term. Background metaphors in law and the perspectives which they create can remain hidden from all but the most sophisticated readers.

When law changes in a substantial way, usually it is the result of change in a background metaphor. A different and more appropriate metaphor may be created by insightful judges or by imaginative legislators. New metaphors in turn provide a new way of looking at old problems and consequently new ways of solving them. To speak of solving legal problems may be too strong. The best we can do with most problems is to recognize the limits created by a previously used metaphor and then either to refine understanding of it (so as to provide another way of organizing facts and circumstances) or to adopt a totally different metaphor. Ultimately, however, either the new or the revised metaphor will show its own strains or limits, and thus the process continues.

METAPHOR IN MEDICINE

Philosophers and historians of medicine are familiar with the influence of background metaphors for understanding disease and illness. Medicine's history is a history of changing metaphors. The classical Hippocratic metaphor of balanced humors produced its own disease categories and dictated its own therapeutic responses. Replacement of the humoral metaphor after about 2000 years with an anatomical/physiological/pathophysiological metaphor began the move toward what we know as modern scientific medicine.[84] The process, however, did not end there.

Today we are faced not just with one mainline metaphor in medicine, but with multiple metaphors vying with one another for dominance. Different medical specialties often use very different metaphors and background assumptions. The genetic revolution was ushered in by the discovery of the structure of DNA and led to the creation of the newest metaphor for disease and therapy. The launching of a genome project to identify genes and their functions gave tremendous impetus to the creation of new specialists, new disease categories, new therapeutic strategies, and yet another way of understanding medicine.

One of the most powerful metaphors in all modern medicine is war. This metaphor has an enormous effect on medical priorities and the practice of both doctors and nurses. It throws light on what we talk about as medicalized dying. Nurses, as well as doctors, will practice under the influence of a set of background assumptions which are pulled together under the metaphor of war.

President Nixon, for example, declared war on cancer back in the 70's. But the war metaphor preceded his use of it. And, when we talk of war in medicine, we mean war to the end, unlimited war, all out war: not war in the sense of just war theory. In the medical context, war is won by defeating death. If a patient dies, that means that we have lost the war.

Listen to our ways of talking about medical practice today. We *battle* disease. We plan *attacks*. We order *batteries* of test. We search for *magic bullets*. Doctors think of themselves as being on the *firing line*, and in the *trenches*. They treat *aggressively*, especially *invading cells*. They take *heroic action* to stimulate the body's *defense system*. The top doctor in the U.S. is a *Surgeon General*.

Older metaphors, however, never disappear totally. Consequently we find ourselves today in a setting in which different metaphors and perspectives co-exist within modern medicine. Alongside different mainline scientific perspectives, older metaphors continue to generate alternative concepts of disease and therapy. Homeopathy has a metaphor (*similes similibus curantur*), and it has not disappeared. Neither has osteopathy or chiropathy. Folk medicine, which preceded all attempts to develop a scientific approach to disease and therapy, not only has not disappeared but today is making a comeback along with its associated spiritualistic metaphors for life and illness. Within each tradition, a metaphor creates disease categories, therapies, and styles of doctor-patient relationships. Along with changes in ways disease is understood are changes in the way doctors and patients interrelate.

INFLUENCE OF METAPHOR ON THE DOCTOR-PATIENT RELATIONSHIP

The humoral metaphor of Hippocratic medicine was linked with a fiduciary metaphor which found expression in an ethical code which has endured 2500 years. In the background of the central ethical obligation not to do harm was the fiduciary metaphor of a healer who could be trusted to do what was best for his patient. The fiduciary metaphor was first expressed paternalistically. Paternalism refers to a doctor-patient relationship patterned upon a parent-child metaphor. It was the reigning metaphor for doctor-patient relations in classical Greece.

The paternalistic metaphor turned out to be much more enduring than subsequent competitors. It survived one change after another in medical science. Today, the paternalistic metaphor is officially rejected but it still influences the way doctors and patients interrelate. Some patients today may complain about being infantilized by their doctors, but most people take for granted a medical version of the parent-child relationship. They don't mind being treated like children when they are in serious pain or when they don't know what to do. In fact they tend to seek out doctors who treat them like a loving parent.

Persons familiar with today's health-care delivery systems recognize the impact of substituting an economic for the classical paternal metaphor. The free market capitalist metaphor has increasing influence on the doctor-patient relationship today. Instead of a patient we now have a health-care consumer, and instead of a doctor we have a person in the business of providing health care. The older power alignments are reversed in this new metaphor. Before, it was the doctor/father who had the power, and he did what was in the best interest of the patient/child without providing disclosure and requesting consent. Where the new metaphor has been installed, it is the consumer or customer who has the power, demands disclosure, makes the choices, even suggests a preferred medication; and the "provider" does all that he/she can to please the "customer."

Before the classical Hippocratic period, the doctor was understood with the help of a priestly metaphor. In the surgical tradition, the doctor was understood and related to more as an engineer. In many of the alternative medicines, the healer is an educator and the patient a student. In legalistic cultures, like the U.S., the doctor-patient relationship is often understood with the aid of a contract metaphor, one consequence of which is to make the doctor vulnerable to lawsuits. (Patients can gain astronomical monetary awards if they have the "right lawyer" to argue their case for contract violation.) The same doctor-patient relationship may be understood metaphorically as a clenched fist or an act of love.

Today's proliferation of metaphors in medicine raises several questions: Are all metaphors equal? How useful, how effective, how respectable is one metaphor compared to another? Does it make any sense for bioethicists and judges and philosophers of medicine to reflect upon the impact of metaphor on ethical standards and legal judgments about medical matters? Is this the right time to re-think the background metaphors operating within medicine? Is there a reigning or enduring metaphor which now should be retained but revised?

THE FIDUCIARY METAPHOR

The paternalistic metaphor lasted for more than thousands of years. Physician duties toward weak and vulnerable patients were considered to be analogous to a father's obligation toward a weak and vulnerable child. The doctor was expected to do what was best for his patient not unlike a father does what is best for his child. Even after this metaphor was altered by patient informed-consent requirements, the fiduciary component of this older metaphor remained.

The term fiduciary comes from the Latin *fiducia*, meaning confidence, and from *fidere*, to trust. A fiduciary relationship is one founded on faith and trust: a relationship in which one person rightly puts confidence, trust, and reliance in the other from whom help is sought: a relationship which requires one to act in the benefit of the other rather than in his or her own benefit.

In the background of most ethical standards and legal judgments about medical matters, one can still find the fiduciary metaphor. The doctor may not be assumed to be a father or to treat his patient like a child, but the doctor is still expected to do what is best for the patient and not what is best for himself or herself. Legislation and judicial judgments as well as medical codes and bioethical standards offer testimony that a doctor is expected to pursue the interests of patients rather than his or her own interests. This is the longstanding background medical metaphor.

Since Hippocrates, the physician has been understood as a professional. The professional metaphor and the fiduciary relationship are linked. The ethical values which dominate in a fiduciary relationship are trust, beneficence, non-maleficence, respect, and now, recently, even truthfulness. What law and ethics attempt to protect and guarantee are these values, which are considered critical for humane medicine in a humane society.

Different metaphors can be shown to have influenced medical practice as well as ethical and judicial judgments. The fiduciary metaphor, however, remains dominant. Long after the official rejection of paternalism, doctors are still expected to be professionals and to fulfill a fiduciary role. The fiduciary metaphor is the background assumption in medical ethics codes, in legal decisions, and in the literature both of bioethics and biolaw.[85]

FIDUCIARY METAPHOR IN SCRIPTURE

The quickest way to get a clear picture of what the fiduciary metaphor means is to look at a familiar parable of Jesus about the untrustworthy steward told by St. Luke in Chapter 16. The story recounts how a dishonest steward took care of himself and pursued his own interests instead of his master's. The steward or manager was supposed to be doing what was beneficial to another (his master). Instead, he did what was for his own benefit. From the story, Jesus draws these conclusions: "If you can trust a man in little things, you can also trust him in greater; while anyone unjust in a slight matter is also unjust in greater. If you cannot be trusted with elusive wealth, who will trust you with lasting? And if you have not been trustworthy with someone else's money, who will give you what is your own? No servant can serve two masters. Either he will hate the one and love the other or be attentive to one and despise the other. You cannot give yourself to God and money."[86]

Applied to the situation of the doctor, Jesus' point is clear. The doctor must serve one master, advance the interest of one person, and do what benefits the patient. If instead she does what benefits herself, then she is untrustworthy. The fiduciary metaphor means that doctors who promise in a public oath to serve the interests of their patients must do just that. They cannot violate that trust by serving themselves or by doing what is best for some third party. Established medical ethical standards and thousands of legal judgments reflect the influence of this one long-standing fiduciary metaphor.

The fiduciary metaphor in medicine does not require that doctors practice without monetary compensation but rather that they do not place their own selfish interests over the interests of weak and needy patients. In the law, doctors are held to higher moral standards because they are fiduciaries. Laws hold doctors accountable for the way they relate to and treat their patients. Where the free market economic metaphor reigns alone, customers are warned to be on guard (*caveat emptor*). This is not the case in medicine, but just such an emerging possibility is the reason why consideration of the metaphor question is so important.

Throughout medical history, ethicists and judges applied the fiduciary metaphor to ever evolving contexts. Today, both are called upon to use even greater insight and creativity in applying it. They have to determine what the fiduciary relationship means in different clinical situations: ordering genetic tests, making diagnoses, giving advice, prescribing, performing procedures, admitting patients to health-care facilities, resolving dilemmas in which respecting one patient's interest inevitably involves doing harm to another.

In the last 25 years the fiduciary metaphor has been extended to assure that doctors will tell patients the truth about their condition and protect their freedom to choose from among real medical options. During that same period the metaphor has come under considerable strain because of changes in the economy of health-care delivery and because an ever-developing medical technology has created dilemmas for which there is no easy solution.

RESPONDING TO FIDUCIARY METAPHOR STRAIN

This traditional metaphor and the perspective which it creates are now under considerable pressure because more and more physicians work for and are paid either by the state or by a capitalistic medical enterprise (HMO). Employers increasingly tell doctors what to do and demand their loyalty. How can doctors be fiduciaries for their patients and at the same time good employees? How can they claim to do what is in their patient's best interest in this new climate? By which metaphor will judges and ethicists evaluate a doctor's behavior: by the classical professional/fiduciary metaphor or by a loyal employee metaphor? This is neither a superficial nor an unimportant issue. The very soul of medicine is at stake. If the doctor is judged by an employee metaphor, this marks the end of the profession of medicine.

The American Medical Association is aware of the importance of this issue and in its codes and ethical opinions continues to hold doctors to their fiduciary obligations. When conflicts of interest develop between doctor and patient, "a physician must exercise medical judgment independently of his own or a third party's financial interest."[87] Any conflicts "must be resolved to the patient's benefit."[88] Conflicts must be understood in terms of a "physician's role as a fiduciary, i.e., a person who by his undertaking has a duty to act primarily for another's benefit."[89] The American College of Physicians in one of its ethics statements made the same point.[90] So have many medical specialty groups in their official ethical statements.

Over and over again in books and journals the primacy of a patient's welfare has been argued by Dr. Edmund Pellegrino.[91] This pioneer in American bioethics is an advocate of maintaining the traditional ethical standards of the medical codes. Patient benefit, for him, is the standard against which both physician interventions and public policies or laws must be measured. According to this ethical standard, the doctor's work cannot be a business. Health cannot be just another commodity. And the relationship between doctor and patient cannot be just a contract. The only ethically defensible role for a physician, he insists, is to be a patient advocate, a patient agent, a patient trustee and a fiduciary of the patient. No social or economic responsibility, for Dr. Pellegrino, can even challenge this first, and indeed sacred, obligation.

Courts and legislatures in the U.S. also insist upon respect for the fiduciary character of a doctor-patient relationship. Courts often talk of doctors as fiduciaries. They require that physicians not abandon patients, that they respect patient confidentiality, that they provide patients with truthful information. We see now, in a beginning phase, applications of the fiduciary metaphor to financial aspects of the doctor-patient relationship. Doctors are forbidden legally and ethically to split fees or to accept kickbacks. Medicare and Medicaid patients cannot be referred to labs which the physician owns or has invested in.

Physicians are not yet held to the same standards as professional economic fiduciaries, but malpractice judgments hold physicians to demanding fiduciary standards in financial matters. The federal government is becoming much more aggressive about false billing. The American Medical Association is increasingly critical of acceptance of gifts from pharmaceutical firms. Despite the complexities created by today's medical delivery systems, the doctor must pursue the patient's interests rather than his or her own or those of an insurance company, pharmaceutical industry, hospital, government, or HMO. But how can this be done with all the pressure on doctors from the opposite direction?

It is one thing to show the over-arching importance of the fiduciary metaphor for physician ethics and its influences on legal statements about physician responsibility toward patients. It is quite another thing to apply this metaphor and its associated presumptions to particular medical situations. Doctors in a particular circumstance may owe loyalty to more than one set of interests. The obvious instance is a doctor's obligation to all members of a community or society when her patient has a disease condition or engages in behavior which threatens public health. Psychiatrists are often faced with just such a dilemma. Increasingly, laws require that doctors report not just AIDS but HIV-infected patients even though such may not be in a particular patient's interest. Threatening infectious diseases have strengthened doctors' commitment to public health and weakened the commitment to a particular patient. This change alters the fiduciary metaphor by recognizing the interests of another person or persons and requiring the doctor to balance these other interests with those of her patient.

Internists and family practitioners now act as gatekeepers for hospitals and HMOs. They control access of patients to specialists. They are required to discourage (sometimes by not disclosing information) expensive treatments which the HMO or insurance company or the government considers marginally beneficial. Conflicts today are everywhere as physicians leave private practice to work for teams, military services, industries, schools, governments, or health-care organizations, all of whom require that physicians recognize other interests and follow organizational policies.[92] Doctors are expected to serve a particular group or a population of persons rather than particular patients who come to them for help. Bioethicists, legislators and judges are challenged to find ways of adjusting the fiduciary metaphor to these new strains, or they have to create an entirely different metaphor, followed by a different ethics and then new laws.

ABORTION AND THE FIDUCIARY METAPHOR

The focus thus far in this chapter has been on different strains on the fiduciary metaphor caused by doctors being employees rather than independent practitioners. Conflict and strains however, may arise between two patients rather than between a patient and an organization or a population of persons (the public or shareholders). One such conflict may be between a mother and her severely defective fetus.

Usually in pregnancy, the physician can be trustworthy and pursue the best medical interests of both mother and fetus. But this is not always possible. If continued fetal life threatens the mother's life, reasonable Natural Law Theory and legal systems using the fiduciary metaphor understand the physician's primary interest to be the mother's life. Even the most pro-life legal systems recognize the inevitability of tragic situations and move away from protection of the fetus to protect the mother. Therapeutic abortion legislation reflects this adjustment. In the case of therapeutic abortions, the fiduciary metaphor is not rejected but modified to prefer one life to another in circumstances of tragic conflict.

Developments in prenatal diagnosis and therapy have increased the instances of maternal/fetal conflict. Now fetal defects and deficiencies can be diagnosed which before did not appear until after birth. The new diagnostic and therapeutic capabilities of perinatology make the fetus into a separate patient. Sometimes, the defects and deficiencies of this new patient can be treated and cured in the womb. Sometimes, however, the fetal abnormalities are so severe as to make life and personal development beyond infancy impossible. Sometimes, a fetal patient is already in the dying process.

Can the classical fiduciary metaphor be modified to address these new problems? The mother's physical life may not be in danger, but the very concept of pregnancy, in the sense of sustaining a developing human being, is strained by the terminal condition of some fetuses. Consequently, the mother's psychic life may be very seriously threatened. Can the fiduciary metaphor be adjusted to address this tragic new situation? If one of the doctor's patients cannot be helped, if treatment of the fetus would be futile, wouldn't the fiduciary metaphor justify shifting medical attention toward doing what is beneficial to the other patient?

Catholic hospital policies often understand the fiduciary metaphor to mean that fetal interests always triumph. The fiduciary metaphor is interpreted to mean that even purely biological fetal life is an absolute value, an interest preferred to any other. This is unreasonable and violates Natural Law thinking especially if the fetus is already in the dying process. Purely biological or vegetative life need not be sustained. Extending a dying process is always wrong and yet seems to be required by such an interpretation. Conservative moral teachings and Catholic hospital policies in effect may require that fatal natural processes be ignored and violated. This first wrong is linked with an additional wrong to a mother, who may be in terrible emotional pain and psychic stress waiting for the tragic delivery.

What if the fetal life is destined only to suffer and never to grow beyond infancy? Would early delivery, or removal of that fetus from the womb violate a doctor's fiduciary relationship to the fetus, if continued existence only extends a painful dying process? To say that fetal removal or early delivery could never be morally acceptable would mean that one narrow interpretation of the fiduciary metaphor is unable to be modified in order to address a tragic situation and in order to prevent multiple harms to another patient. Such intransigence is irrational and a violation of the natural order. It also deviates from traditional Catholic moral teachings reflected in the Church's two greatest theologians. Both St. Augustine (*De Ordine*) and St. Thomas Aquinas (*Summa, Secunda Secundae*, Q 10, Article 11) insisted that evil happens, and sometimes authorities have to permit evil because otherwise greater evil will incur.

Because making fetal life an absolute is so absurd and does so much harm, many advocacy groups push to eliminate the fiduciary metaphor altogether and replace it with a totally different one. They would prefer to consider the fetus a non-entity, or a body part, or a foreign growth, which mothers can dispense with at will. These alternative metaphors are inappropriate, dangerous and certainly disrespectful of human life. Much more appropriate would be a refinement or modification of the classical metaphor so as to be able to address tragic medical situations. The modification or improved interpretation is actually very slight: the doctor would be obliged to pursue the interests of both fetus and mother, but when the fetus is dying, or cannot develop, or faces only suffering, then a mother's interest in terminating a failed process of fetal development is given priority. Choosing fetal death in tragic situations does not make it a good but rather a lesser evil. Some pro-life advocates ignore the reality on which ethics is grounded.

Total abandonment of the fiduciary metaphor for both mother and fetus would lead to widespread suffering and death. It would have negative consequences for women and fetuses and for any society which has an interest in the protection of human life. This traditional metaphor continues to hold physicians accountable for doing what is best for patients and for respecting patient life. The same is true of traditional Catholic moral teachings.

Couldn't a Catholic hospital policy define more sharply the limits of the fiduciary relationship to a fetus who is beyond medical help, and destined to suffer, and never to develop? Couldn't a more sharply defined metaphor create policies which shift physician beneficence toward a suffering mother when the fetus is beyond beneficence? Couldn't hospital policies address the inevitability of tragic situations in pregnancies and look at the Catholic tradition for handling tragedies? Couldn't Catholic hospital policies justify removal of a fetus from a process which is fatally flawed, something which nature itself ordinarily does in such circumstances through spontaneous abortion? When the fetus is dying or faces terrible suffering and can never develop to personhood, couldn't doctors do what nature usually does: i.e., stop the fatally flawed process? In such situations, couldn't hospital policies support imitation of nature, improvement of a natural process, relief for the fetus, and respect for interests of the mother? A Catholic perspective based on Natural Law and carefully analyzed objective reality would answer in the affirmative.

Abortion laws and hospital policies have to balance the interests of a mother and fetus when they become tragically locked in conflict. Undivided or absolute loyalty to one or the other would be a violation of nature and a distorted understanding of the fiduciary metaphor. The fact that physicians have a fiduciary relationship means that it must be adequate to conflicts of interest which may develop. Conflicts of interests are not a reason to abandon the classical metaphor. Rather, they are a call to review medical dilemmas to make sure that physician and hospital accountability are not lost, respect is shown to both patients, and a nature-based sanctity of life principle is safeguarded. Conflicts and tragic dilemmas are a call to make sure that death is not prolonged, that harm is not done without compensating benefit, that respect for a mother's psyche and conscience is supported by hospital policy. Only when the fiduciary metaphor is interpreted so as to place all interests on one party and to deny the interests of the other, do both anti-abortion legislation and hospital policies fail.

When abortion is made just a matter of personal choice at any time, it requires procedures (like partial birth abortions) which are so horrific that no doctor could stomach them for long. Abortion as pure choice for any reason creates devastating medical problems for women. It creates a fetal holocaust because it denies the nature-based value of the developing life. It violates an ethics based on nature itself. On the other hand, restrictive legislation which absolutizes fetal interests and prohibits every termination of pregnancy no matter what the circumstances, creates a holocaust of women caught in tragic conflicts with their fetuses. It violates nature and reasonableness and humane medical practice.

Pregnancy is not the only situation in which professional physicians bound by fiduciary obligations have to take into account the interests of other parties. In contemporary health-care systems, physicians have to do so constantly. In most places, however, the fiduciary metaphor has been adjusted and re-interpreted rather than abandoned. Laws and hospital policies continue to hold physicians accountable to patients even though they also have obligations to other interests. Public health situations, for example, do not require an abandonment of the patient, but refine the extent of patient interests that require respect. Modifications of the fiduciary metaphor are controlled by careful restrictive limits. The same adjustments could be made for conflict-of- interest situations in pregnancy.

The argument being made in this chapter is a very general one. I am trying to show that behind public policies and hospital policies are metaphors which steer decision-making in a certain direction. I want to argue that such paradigmatic metaphors need to be made explicit, reflected upon, and constantly refined. If certain metaphors have endured thousands of years, it is because they are important. They guarantee important societal values. Only critical reflection and continued refinement of these metaphors can guarantee their continued appropriateness.

The fiduciary and professional metaphors used in medical matters are so old as to qualify as venerable. Modern health-care delivery systems, both socialistic and capitalistic, threaten these metaphors. Physicians may very well become workers, ruled by what their employer dictates rather than by their own independent judgments about what is beneficial to sick and vulnerable people. Maybe another metaphor will take the place of this classical one. Until a better metaphor appears, however, I think the classical one deserves commitment and our reflective attention. A liberal Catholic bioethics would suggest preserving this metaphor by adjusting it. In this perspective, the best way to guarantee that the classical metaphor will fail is to refuse to refine it and to leave everything the way it is.

MODIFYING ABORTION POLICIES

Ethical theorizing about the nature of the doctor-patient relationship or background metaphors or the sanctity of life principle, at some point has to be applied in particular situations through concrete rules. Abstract theories and principles must be translated into practical policies which guide professional decision-making. The following policy suggestions or tentative policy formulae are examples of ethical practices which might be modified. Anencephaly is one specific example of a fatally-flawed fetus, but there are others.

1. The attending physician is responsible for determining when a confident diagnosis of anencephaly can be made. If reasonable doubt exists after alpha-fetoprotein and sonography, ordinary support of the fetus is required. Consultation with physician specialists other than the attending obstetrician is required to confirm a doubtful diagnosis.
2. Once a firm diagnosis is made, the pregnant woman must be fully informed of the fetus' condition and the prognosis. She is the primary decision maker, and determination cannot be made without her informed consent. It is to be expected that different women will have different responses to this tragic situation based on their religious and cultural values and on the support they receive from family and medical staff. Staff members usually provide counsel, support, and advice, and give each woman time to come to a decision.
3. If a woman insists upon not bringing the anencephalic fetus to term, and if her decision is a free, competent and informed choice, it may be respected by the staff.
4. Abortion of the anencephalic pregnancy is appropriate when the diagnosis is firm, and the woman makes a free choice. If doubts persist about the diagnosis or about the free and informed quality of the woman's decision, interruption of pregnancy should be postponed.
5. It takes a certain amount of time to diagnose anencephaly. After the diagnosis, the mother may require interruption of pregnancy, or she may decide to wait. Given the burden to the mother and the absence of either burden or benefit to the fetus, requiring a mother to wait until viability (on the remote possibility of misdiagnosis) is morally unjustifiable.
6. After medical factors are clarified and the woman's decision is made, induction procedures should follow accepted medical norms. The health and well-being of the woman as well as her capacity for future pregnancies are the major medical goals.

7. A woman's decision to carry an anencephalic fetus to full term will be respected by the staff even though carrying an anencephalic fetus increases the mother's risk. Interruption of pregnancy is permitted only if the woman is either burdened or harmed by continuing the pregnancy and considers continuing the pregnancy to be unreasonable.
8. Medical and nursing staff in hospitals who participate in the pre-term delivery of an anencephalic pregnancy should be given careful ethical education in order to avoid misunderstanding or scandal because of what on the surface may appear to be a violation of traditional respect for human life.
9. Ethics committee members should be made available to any woman facing a decision about her anencephalic fetus. If full review by the committee is not feasible, a smaller sub-committee may be available for consultation. The committee should address uncertainties, provide explanations of the hospital policy, and mediate any misunderstandings about anticipated procedures.
10. An anencephalic infant who is delivered alive will receive humane supportive care. No treatment will be provided either in the form of mechanical respiration or medical nutrition and hydration.
11. A live-born anencephalic infant is not brain dead as long as there is brain-stem function.[93]
12. Such an infant must be considered alive but in the dying process.
13. Health-care institutions should be sensitive to the need in our nation for more infant hearts and kidneys and infant livers than are available. The death of 40 to 70 percent of children (under two years) on transplant waiting lists is a particularly serious statistic.[94] But until transplant technology from anencephalics is perfected (for example, donor-recipient match) and the ethical questions surrounding the determination of death are settled, institutions are restrained from using live anencephalics as organ sources.[95] [96] [97] [98] [99]

DEATH

X. Aging and Dying

INTRODUCTION

The idea of writing a chapter on aging for this book seems particularly appropriate. I'm at the right age to address this topic. As I did the preparatory reading and research, I had an unusual experience. For the first time in my life, I felt like an expert. I could judge what people were saying about aging by my own experience. Besides facing aging myself, I work mainly with rural hospitals and nursing homes and hospice organizations where most of the patients are elderly. Both in my personal life and in my work, I am surrounded by the realities of aging and dying.

Aging and dying are certainly connected with a more humane medicine. All the losses which define illness are present to a degree in aging. Consequently, a humane medicine must first be sensitive to the aging experience and then must be committed to meeting the needs of aging persons. An understanding of aging is important for a practicing physician and for the doctor/patient relationship, but it is important as well for the institution of medicine. Developing a strategy for handling aging is one issue which the doctors and the health-care systems of every culture have to face. We all age, we all die, and we all need humane medical care in the process.

In his recent book entitled Dawn Gray[100] on the economics and politics of aging, Peter Peterson used a gripping metaphor to describe the challenge of an increasing aging population. Global aging, he said, is like a massive iceberg which very well could destroy even the most powerful economic vessels in the world. The aging population worldwide, according to this author, threatens human survival and is the most important challenge we face in the 21st century. His iceberg metaphor highlights the danger of global aging, the fact that all human beings are exposed to this danger,[101] and the fact that aging is a challenge which doctors and the medical establishment have to take seriously.

THE SENIOR PERIOD TODAY

The modern industrial system with its pension plans and a specified time to stop working has created and defined the new period in contemporary culture which we call retirement. As the numbers of retirees grew, older people achieved a separate social status. The elderly became seniors. Now we have a senior period, with senior people and senior moments and senior discounts and senior cruises and senior housing. A new nursing-care industry has been developed for handling the worst medical problems of this population. The government has stepped in to defray some of the medical costs with Medicare and Medicaid. A whole new academic discipline (gerontology) has developed to try to understand this period.

Spain's greatest 20th century philosopher Jose Luis Aranguren sub-divided the senior period into *la tercera edad*, the healthy senior period, and *la cuarta edad*, the disabled or infirm senior period. He used new concepts and categories for one and the other stage: *viejos*, the old, *senesientes*, the senescent, *ancianos*, the aged, *senectos*, the decrepit, *longevos*, the long lived, *valetudinarios*, the infirm or sick. Often, he made a distinction between knowing that one is a senior and feeling old.[102]

At the fringes of every aging experience is increasing pressure from the reality of death. Many of the senior activities in American culture come over as distractions from or even denials of this reality. Death in the U.S. often is treated as a taboo topic. Sooner or later, however, death and questions about how to die, force themselves into consideration. Aging anticipates something else, and that something else is death. Death is a part of the aging experience that cannot be ignored, no matter what the cultural peculiarities of the newly designed period.

As people age, ignoring death becomes more and more problematic. What once may have been conveniently ignored inevitably re-enters the picture. At first it may be a vague "deadline." The invisible referee in the game of life at some point will call "time out." And yet, even people in the senior period, glance at death only to look away and move to the next distraction.

MEANING AND OLD AGE

If death is not given due attention in our culture, neither is the issue of the meaning in old age. Modern industries and government programs have helped to define the senior period, but they have done little to provide it with meaning. What is the meaning of life when we retire, or when we get old, or when we become decrepit, or when we get close to death? These questions do not get much attention despite the number of retired persons who slip into depression once the meanings linked to their work and parenting are lost.

Industrialists and legislators created the senior period, but in doing so they created some moral and existential problems. Industrialization linked with Protestantism created what we call the work ethic. It is an ethic in the sense of inner attitudes or dispositions. The work ethic linked the meaning in life, even eternal salvation, with an inner commitment to work. A work commitment became the essential inner disposition. It was the key to wealth in this life, and then salvation in the next.

In pre-industrial society a different ethic operated. Different inner attitudes or dispositions or background personality traits provided moral direction and meaning. The theological virtues (faith, hope, and charity) gave meaning to life. The cardinal virtues (prudence, justice, fortitude, and temperance) gave moral direction. In the pre-industrial society, many good people never worked. Aristocrats didn't work. Contemplatives didn't work. But their lives had meaning and they were respected members of society.

Following industrialization and unionization of workers, working conditions and public health infrastructures gradually improved. Consequently, working people started living longer. Now we have a large segment of the population retired from work. The majority of these people, however, were formed in the work ethic. Understandably, many run into problems when they stop working and enter the senior period.

Some retired people have money and have made a transition from membership in the working class to membership in a consumer society. Shopping now gives meaning to many lives. ("I shop; therefore, I am.") It gives meaning because people become identified with the things which they buy. In our commercial culture, things give meaning. Shopping is important because "I am the things which I purchase." Some people would not be caught dead wearing something from a competing brand. But these commercial identities and meanings create an existential superficiality. Personalities are as passing as their purchases and their new products.

In addition, elderly persons cannot consume the way they once did. And things gradually cease to provide real identity and meaning. Even when shopping and things give meaning, the meaning is superficial, and people are not happy. The new secular meaning system fails to satisfy most seniors. Extending longevity and creating pensions for seniors has not been accompanied by a culturally rich and a socially shared vision of the meaning of being old. Rather, current secular developments have created a moral and existential problem for seniors.

This situation and the challenges it creates was captured in a poem by a retired man in Galveston, Texas. Thomas Cole, a bioethicist and medical humanist at the Institute for Medical Humanities in Galveston, Texas, organized a project to promote emotional growth among elderly people by the use of autobiography. Mr. Bob Burdett captured the point I am making in a short poem entitled *What Now?*

Retired
Off the treadmill.
Out of the rat race.
Affairs in order.
Paper read.
Bills paid.
Laundry washed and folded-
It'll only take a minute to put it away.
Enough time for everything-
Then some.
Too much of a good thing.
Too many crossword puzzles.
Too many naps.

Errands I used to do on the way from Point A to Point B have become major events.

Is this what it's supposed to be like?
I'm retired on insufficient data.
Now I'm expected to live another 25 years.
Almost 10,000 days.
A slow death.
Travel?
Romantic involvement?
Don't think so.
Need to do something
Productive,
Significant,
Meaningful,
Gratifying.
Like re-invent myself.
Start all over.

PAGAN CULTURES

The classical cultures of Greece and Rome had theories of aging which provided meaning by dividing life into specified stages from birth to death. Stages were numbered and arranged in ascending and descending order. Certain years were especially important because the organism was believed to be ordered in seven-year cycles.[103] Seven was the age of reason, 14 the beginning of adolescence, 21 the beginning of adulthood. Between 28 (4x7) and 49 (7x7) the adult lives and develops. After 49, the decline starts. At 7x10, life is expected to be over. Both aging and death were understood to be a part of the natural order.

The seven-year cycles correlated with inner attitudes and expected behavior. There were some differences of opinion about behavior in the final period. Aristotle thought that in old age, a person's normal failings were magnified and physical decline depressed the human spirit. Plato urged the elderly to enjoy the time free of passion. Cicero too thought that old men were relieved of the drive for sexual pleasure. This relief in turn provided the climate for wisdom.

Greeks and Romans gave meaning to aging through their system of natural stages and through submission to the natural limits embedded in the stages. Cicero made the point that "the burden of age is lighter for those who feel respected and loved by the young." (De Senectute, 8, 26). Classical literature, however, is also full of ridiculous characters who behave inappropriately for their age. Old men who lusted after young women were the butt of jokes. Only old people who behaved properly for their age received respect from younger members in their communities.

RELIGIOUS CULTURES

In religious cultures which preceded our secularized industrial age, the lives of persons who made it into old age were full of meaning. How a person aged would determine how he or she would spend eternity. Even ordinary and pedestrian experiences had meaning; indeed, they had eternal significance. The realities of this life were connected to the possibility of yet another life beyond this one. In religious culture all of life is understood as a journey. Aging means approaching the end of the journey. Death is a passage to yet another life.

Old age had a built-in respectfulness in Jewish culture. A long life was seen as a sign of God's favor (Genesis 11:10-32). In Hebrew Scripture, God often chose old people for mighty deeds. The elders in Jewish culture were given prestigious community tasks which gave them respect and dignity.

The same system is reflected in Christian Scripture and later in Christian culture. The aged Simeon and Anna recognized Jesus as the anointed one when he was brought to the temple as an infant. This was fulfillment in their old age. (Luke 2:29). The elderly Nicodemus was praised for his courage in helping with the burial of Jesus (John 19:38). St. Paul in the Epistle to Titus (2:2-5) addresses the issues of meaning for the elderly and their social roles. "Bid the older men be temperate, serious, sensible, sound in faith, in love and in patience. Bid the older women likewise to live in a way appropriate to believers…; they are to teach what is good and so train the younger women to love their husbands and children."

Job showed how to handle the worst losses imaginable in aging. Jesus, too, suffered through tough experiences of loss: first the anticipation of loss of his life in the agony in the garden (Lord, take this cup away from me…Mark 14, 36.); then, the total sense of loss on the cross (My God, my God, why have you forsaken me. Mark 16, 34). Jesus and Job can help old religious persons to find meaning and to endure their losses without falling into despair. In both the Jewish and Christian cultures, old age is a time to face loss and to bring life to fulfillment. In Judeo-Christianity, it is the expectation of what follows this life that gives meaning and the courage to face losses. In light of eternal life, this life becomes a short path along which we walk in order to mature morally and prepare for what will follow.

By contrast, in the contemporary American secular culture, the senior period is expected to be a stage of continued and unlimited development. Death is not even part of the discussion. But such denial collapses either gradually or suddenly into illness and disability. At some point the elderly are forced to face both death and absence of meaning. Consequently, aging and dying are frequently plagued with depression.

No matter what one's culture, aging always involves increasing physiological and psychological vulnerability and the increased probability of death. Aging and dying are physically and conceptually linked. As we age, we have to ask how we should age and then at some point how we should die. We have to ask about the meaning of aging and dying. For human beings, aging and dying cry out for reasons and meanings. Here the Church can and should enter with programs designed especially for older people.

FINDING MEANING IN MEDICINE

One frequent source of meaning for contemporary persons is medicine. Doctors are for many people today what priests were for people during more religious periods. Health is equivalent to salvation. Disease and disability are hell.

Whether or not aging is a disease depends upon one's philosophy of medicine. Certainly there are biological and psychological alterations as years go by, but not all philosophers of medicine would classify these as disease. Medical interventions can often modify these alterations. Physicians have a large arsenal of interventions for age-related changes in the organism. Doctors provide meaning in old age by creating an important resource for elderly persons and by keeping their worst fears in check. Down the line, medicine promises remedies for memory problems, osteoporosis, vision loss, and the control or elimination of chronic illness associated with aging.

Doctors who appear on tv and write daily columns in the newspaper frequently talk about aging. They tell us about the molecular and cellular damage which cause aging. They tell us about the organic and environmental factors which are involved: radiation, toxins, oxygen radicals, stress hormones, modified lipids and proteins, cytokines, glucose, etc. All of the above are subject to medical interventions. Currently, a great deal of medical research is focused on cellular and genetic manipulation which has more than doubled longevity in certain worms and animals. The possibility of making spare human parts from stem cells adds even greater possibilities for extending life, first in animals but ultimately in human beings.

Environmentally caused pressures on life, certainly, can be relieved. Harmful environments can be changed for the better. Life-style practices like lack of exercise and poor nutrition can be improved. People can stop smoking and lose weight. These alterations can extend life span by as much as 50% without recourse to any of the new medical interventions. Doctors and pharmaceutical firms continually endorse these recommendations.

Recent pharmacological discoveries promise to eliminate or at least to reduce the effects of life-shortening diseases. Vitamins and food additives like anti-oxidants can prevent some of the damage from products of cellular metabolism. Effective cures joined with enhancing vitamins and food supplements can postpone aging and add many years of life. Life-enhancing hormone replacements like estrogen can also make a big difference in aging for women.

At the beginning of the 20th century, the average life expectancy in the U.S. was in the 40's. As we begin the 21st century the average life span is in the 70's and 80's. The declines that create aging and move every human being toward death are undergoing modification by medical interventions and public-health improvements. These declines tend to be more precipitous in developing countries and less so in the developed world. Physicians in first world countries have more control over infectious and parasitic diseases. Public-health systems have had the greatest influence on aging statistics.

A small percentage of persons survived into old age in every culture and every historical period. What has changed in the past century is the greater number of those who become old before they die. Aging, in other words, has become the experience of an ever-expanding number of persons. Whether the increases in longevity can continue is a matter of controversy. Most experts today agree that the 30-40 year increases in longevity during the past century will not be repeated in the next, but further extension in life span is anticipated.

Persons over 85 are the fastest growing group in the U.S. population. They are 21 times more numerous then they were only 100 years ago. Persons over 65 are 8 times more numerous than at the turn of the 20th century. There are 68 men per 100 women over 65. Early in the 20th century, it was the younger population that was saved from early death by medicine in the form of public-health improvements and vaccinations. One challenge for medicine in the 21st century is to save the aging population from vascular disease and cancer. A second and more important challenge will be to help people with aging and then with dying.

Even with all the attention which medicine has given to aging, still the issue of meaning or lack of meaning threatens the elderly population. Extending life span and curing disease has not solved the problem of meaning. Does it make any sense to extend life span when the added years are devoid of meaning? Is it right to spend scarce resources to extend meaningless existence? Wouldn't it be ethically preferable to design strategies for more humane aging and dying? Try as we may to push away the link between aging and dying, it will not budge. We cannot talk about the meaning of aging without addressing the meaning of dying. And if there is no shared view of the meaning of aging and dying, then despite all the medical advances, old age and proximity of death are going to be plagued by all manner of unmet needs, but especially by the scourge of depression.

DEPRESSION IN OLD AGE

To despair or not to despair, that is the question elderly people face. Even for persons who enjoy good health and are surrounded by family, being old means being lonely. Loneliness involves a sense of a loss, and depression is failure in handling loss.

Human identity is a construct, and the building blocks of this construct are social roles and relationships. A loss of either can shake one's identity and create the conditions for depression. If my family is gone and I have no social role, who am I?

Loneliness is more likely to cause depression when compounded by dependency. It is even more likely to cause depression if poverty or the loss of one's home and the necessity of living in someone else's space is added.

Loss is everywhere in the last stage of life. For the infirm elderly, old age means physical loss: sight, hearing, mobility, and continence. The last loss is the most embarrassing. Loss of mobility means that someone else has to go to the store and has to take care of everyday needs. Sight and hearing loss close the old person in on him or herself. All these physical losses are tough to manage.

Old age means mental loss as well, especially reasoning ability and remembering. It means not to be able to find your glasses or to keep things in order. It may even mean the loss of security and independence. Who wouldn't be depressed under such conditions?

If there is any doubt about the pervasiveness of loss and depression in old age, listen to some of the lines spoken by elderly characters in Shakespeare's plays:

> I shall despair. There is no creature loves me,
> And if I die no soul will pity me
> And wherefore should they, since that I myself
> Find no pity to myself.
> Richard III, V, iii, 200-203
> Here I stand your slave
> A poor, infirm, weak and despised old man.
> King Lear, III iv, 19
> A poor old man,
> As full of grief as age.
> King Lear, II iv, 271.
> I have lived long enough. My way of life
> Is fallen into the sear, the yellow leaf,
> And that which should accompany old age,
> As honor, love, obedience, troops of friends,
> I must not look to have.
> Macbeth, V iii, 22
> Grief makes one hour ten.
> Richard II, I, iii, 260

Even if the connection between old age, loss and depression is not a recent or a new phenomenon, the question today is how to help the growing population of elderly to maintain meaning and hope. In the U.S. secular culture today, we try to keep the elderly distracted. We take them on trips.

We teach them how to play cards. We start exercise classes and dance classes. We take shopping and buying tours. Rarely do we honestly confront the experiences of loss. Rarely do we try to make sense of these experiences. Rarely is depression in the elderly either accurately diagnosed or adequately treated. Focus is concentrated on the external aspects of life. The internal or existential or psychiatric aspects are not addressed. This is certainly one of the ethical problems associated with aging.[104]

AN ETHICS OF AGING

Any ethics of aging and dying must address the problems of loss and depression. Ethics must grapple with the issue of meaning in the aging experience (an existential ethics). A humane ethics must be grounded on Nature and must establish limits based on the structure of human reality.

Then humane ethics must provide a response to the many social changes brought about by the big increases in the aging population. (social ethics).

The elderly 12% of population consume 33% of medical expenditures. Increases in the number of older people inevitably involve increases in medical costs. Pressure is already being felt to shift spending away from the elderly toward younger people who have to bear most of the health-care costs. Increases in the number of elderly persons place health-care burdens on younger people. More and more frequently, younger family members express resentment over the sacrifices which they are expected to make to care for elderly and needy parents and grandparents. A reasonable ethics has to address this issue.

Ethical questions are being raised about the limits of associative or relational obligations. And questions are being raised about whether resentful younger family members can act as surrogate decision makers for elderly members. Serious consideration is now being given to rationing certain life-saving medical interventions based on age: e.g., no renal transplants or open-heart surgery after a certain age.[105] The most powerful exponent of the idea that medical technologies have to be limited is Daniel Callahan, the founder of The Hastings Center.

The elderly population also has its advocates and lobbyists, and they have different ethical concerns. Besides attention to mental and emotional problems among the elderly, they insist that increased attention be paid to traditionally ignored medical problems. People over 65 are finally being included in research trials which address their most prevalent conditions (e.g., heart disease, cancer, diabetes, hypertension). It was wrong to exclude this population from trials of medications which might improve their condition.

The expanding population in *la cuarta edad*, the disabled elderly, translates into an expanding population of disabled persons who need some form of long-term care: day care, in-home services, convalescent home care, intermediate and skilled nursing home care. Seventy percent of persons over 85 need some kind of help in order to live from day to day. And this help can be very expensive. It can quickly exhaust an elderly person's savings. As family size goes down in developed nations, more and more pressure is exerted on fewer and fewer younger, (usually female) family members. This is a major ethical concern.

Outside help for infirm elderly is commonly provided by informal caregivers. But this kind of help is in short supply in government programs. For most elderly people who need care, family and friends have to provide the help. Government programs for the elderly tend to be for acute care rather then long-term care. They cover a broken hip, but not the inability to feed oneself.

An infirm elderly person may need expensive drugs, a prosthesis, some structural changes in his or her home or apartment, a rehabilitation program, meals, shopping assistance, etc. Human beings need help of some sort at every period of life, from birth to death. The very well-off elderly can avail themselves of expensive assisted-living arrangements, but most elderly need some form of social or government program. Otherwise, they have to depend on the charity of family and friends. But what kind of program, for whom, at what cost? These are the critical elements in any decent social ethics for the elderly. And where does a more humane medicine enter the picture?

Some elderly people live for many years with chronic problems and needing chronic medical care. Others live relatively healthy lives until some acute incident throws them into need: a stroke, a fall, a heart attack, a cancer, and the death of a spouse. Suddenly, functional abilities decline, and help from other persons is needed. If the other persons are not family members or friends, they are likely to be non-professionals, poorly paid, and unreliable because they themselves are needy. This is a problem.

The more humane medicine and Catholic bioethics of aging being suggested by the above considerations takes the form of agreed-upon public policies which address the most pressing needs of the infirm elderly. If care is provided by non-professional outside sources, these should be licensed, and the help they provide should be monitored. If family or friends provide care, the limits of what these persons can provide should be stipulated. Family and friends should not be overwhelmed by the care giving. Policies should not deny support to family care givers. Finally, professional caregivers supported by public policies should maintain communication with family caregivers and help them when necessary. The ideal goals of a social ethics are easier to identify than the concrete policies required to fulfill the goals.

A POSITIVE CULTURE OF AGING AND DYING

One response to the campaign to convert suicide into a solution for problems of aging and dying would be a campaign to extend mental health care to the elderly and to make aging more meaningful and dying more humane. Instead of retirement and aging being a loss of work and meaning and therefore a depressive period, it can become a time for spiritual development. Instead of being a time of self-centeredness it could be promoted as a time of altruism and generosity. The senior period could be conceived of and promoted as a time for the development of age appropriate positive attitudes and dispositions rather than depression and isolation.

The senior period could be time for intellectual development as well. The experiences of life could be thought through, and lessons drawn which then could be communicated to those who ask for advice. Elderly people could spend time reading and reflecting and writing down for others some of the wisdom they have acquired. Instead of spending time shopping, elderly people could spend time in libraries doing their reading and reflecting, then in churches and community centers where people who need help could find them. Mainstream secular culture provides no free and easily accessible help for people who may need direction and advice.

Elderly people could focus, too, on developing a more sane and more healthy rhythm to life. They could provide a needed alternative to the hyperactivity characteristic of persons in today's work world: the computer work at home and on vacation, the phone communication while walking on the street, working during meals, etc. Elderly culture could be more spiritual and less material, more peaceful and less agitated, less involved in cyberspace and more invested in simple pleasures.

People who live in consumer cultures wind up collecting things. By the time they reach the senior periods, many people find themselves with houses full of things. The senior period could be understood as a time for giving things away. Instead of being identified with things, seniors could start a process of disengagement. They could teach themselves, and in the process teach younger people by their example, that life is not about collecting things, but about understanding, and relationships, and becoming certain types of persons. Some of the collected things could help the disadvantaged in other cultures.

For persons whose lives are absorbed in work and frantic activity, seniors could become an example of life that has time for others. The only persons in today's culture who make time for others are professionals who charge big fees for a 45-minute visit. And the visits are on their schedule, not yours. Elderly people could try to make themselves available anytime. Young people especially could always stop in and be welcomed and be listened to. Even though elderly persons might feel inadequate to give advice, the fact that they were welcoming and available and willing to listen would make an important contribution to the whole culture.

One salient characteristic of people in contemporary secular culture is self promotion. People will do anything to have their 15 minutes in the spotlight. Self promotion is one aspect of whatever other product or service that is also being promoted. Instead of being focused on oneself, seniors can show the attractiveness of just the opposite way of being. The senior could show life that one can live without constantly making reference to oneself. Seniors can show that one can live and be attractive precisely by not being self-promotive but rather by giving space to others.

Finally the senior period could be redefined as a time of courage: the courage required to keep despair and depression at a distance. Aging inevitably involves physical and mental decline just as inevitably it involves the loss of loved ones. Consequently, *la tercera y cuarta edad* (the third and fourth age) are often plagued by depression. Courage can keep despair and depression at a distance. Those who show such courage could be held up as models for persons at every age.

Every life has its losses, and loneliness is something experienced by everyone. But the loneliness of an elderly person who has lost friends and family is profound. Instead of withdrawing, the courageous senior can resist regression. He or she can stay in contact with other people, can help younger ones, and can share wisdom with the immature. The courageous senior who resists despair and depression can become the critically needed proof that inner strength is real and that despair and depression are not the fate of human life.

Extreme advocates of legalized suicide do not make exceptions for persons who are depressed. Either they are ignorant of the informed consent requirement for any such medical assistance or ignorant of the degree to which depression impairs a person's cognitive capacities and makes an informed and free consent impossible. Options are so narrowed by depression that rather than being freely chosen, death is seen as the only option. When mental processes are paralyzed, and options restricted, there can be no informed consent. Hopelessness and unbearable pain are not background conditions for a free and informed decision to take one's life.[106]

CONCLUSION

The specific aid and support which is ethically required for the infirm elderly is not clear. Socio-political movements to establish long-term care, if they exist, have not yet reached extensive public recognition. Specific, practical, affordable and democratically acceptable policies have not yet been developed which articulate precisely what help humane communities should provide for needy old persons. Standards for determining need in other areas of life may not apply to the situations of elderly people.

Abstract ethical principles like equity and Natural Law-based concepts like the common good obviously are important, but these abstract ethical standards do not by themselves show what practical concrete policies are appropriate. Sound, effective, affordable, concrete policies for long-term care are more of a challenge than easily identifiable ethical abstractions. Good policies meet real needs with basic benefits.

Natural Law reasoning can arrive at abstract nature based rights but the exact content of *basic rights of elderly people* is not yet clear. The *right to health care*, for example, is a credible concept, but like the concepts of *equity* and *common good*, it is still an abstraction. Its application to broad groups of persons requires the involvement of social sciences, medical sciences, economic sciences, and organizations like the World Health Organization. Even after issues of macroallocation of resources for needy elderly have been settled, still the basic rights concept requires practical application in the sense of microallocation, i.e., which particular needs of individual elderly persons can be met. The frail elderly as we have seen have both physical and mental needs. It makes sense to start with the basic Natural Law perspective and move from there to a *right to health care for the elderly* concept as the foundations of a humane medicine. The remaining challenge is to apply these concepts to concrete individual situations. Developing a balanced and workable ethics of aging is a task which has hardly begun.

Doctors treating elderly patients have to take their existential conditions into consideration. They have to address issues of pain and suffering. If religious meaning is gone, some kind of meaning system has to be invented to take its place. One project that might generate cooperation among split denominations in the Christian community would be an attempt to restore religious meaning to ever-increasing numbers of aging people who have lost their faith. Liberal and conservative perspectives might both have a place in such a project. Instead of competing with one another or continuing to argue about doctrinal and ethical issues, liberals and conservatives could come together to keep those who are aging and dying.

Finally, humane medical practice has to extend to the moment of death. Humane medicine has to do everything possible to help the aged, and indeed all patients who are dying, to die well. Since the beginning of modern medicine, the focus both of patients and of doctors has been on curing disease and longer life. Gradually doctors and patients have begun to change. States such as Michigan that rejected Jack Kevorkian's vision of dying, have taken a leadership position in designing an alternative plan for humane dying.

All aging persons, especially elderly dying patients, need information about hospice, a program that stands for humane dying. It has to be easier for doctors to prescribe pain medication for the elderly dying. Hospice is an advocate for this change. Hospice stands for providing humane care including pain relief, social services, attention to religious needs, family support, etc.

Doctors now are being retrained to help patients die more humanely. They are being encouraged to talk to patients about dying well instead of moving away silently when their treatments no longer work. Access to more humane dying is being covered in private and government health plans. Campaigns are being organized to inform patients and families about what is now available, instead of the inhumane continuation of aggressive therapies to the end.

Achieving a more humane medicine requires cooperative efforts by government, non-governmental programs like hospice, hospitals, schools and churches. To alter a now famous phrase, it takes a village to make aging and dying more humane.

XI. Euthanasia and Physician-Assisted Suicide

INTRODUCTION: MEANING OF THE WORDS

If a liberal Catholic, Natural Law-based ethics avoids fundamentalism and the illusion of changelessness, it is because it insists upon looking at moral teachings in an historical perspective. Historical understanding puts official moral teachings in context.[107] It may open teachings and policies to change in light of new contexts or it may reinforce and strengthen traditional stands. There are books and articles which look at ethical issues like euthanasia and suicide in historical perspective, but they have not attracted much public attention.[108] Before the current cultural debate about legalization of assisted suicide and euthanasia, it was not a pressing public issue.

The first point to be made about euthanasia is that the term is relatively new in the present meaning. Its etymology is simple. *Eu* in Greek means well or good. *Thanatos* means death. *Eu Thanatos* means a *good death* or an *easy death*. The etymology carries no suggestion about how the good death is achieved. The Oxford dictionary cites 1646 as the date of the first English use of the term, and for the next couple centuries, the word did not mean a shortening of life or a taking of life. It referred to the state of a person's mind at the time of death. A tranquil state at death would be referred to as euthanasia. This could be brought about by non-medical persons: clergy, friends, and family. Only at the end of the 19th century did euthanasia begin to acquire its present meaning.

Suicide, like euthanasia is a 17th century term. Its etymology is Latin rather than Greek. *Sui* in Latin means *of himself*. *Cide* derives from the Latin verb *to kill - occidere*. When the term suicide was introduced into English, it had a more morally neutral meaning. Before its introduction, self-killing was called murdering oneself. Over the years the term suicide acquired a strongly negative moral connotation, but efforts are underway again to make the term more morally neutral.

Some very general things about the death of human beings are well known. For example, throughout human history death was never just a physical event. Some theory of death had to be applied to determine when death actually took place. Then the death event was followed by rituals of some sort which human beings everywhere have created to help them manage this loss. One sure proof that a skeletal remains is that of a human being turns out to be its proximity to relics of death rituals. Definitions of death and funeral rites are among the distinguishing characteristics of human animals.[109]

Anthropologists believe that Neanderthals were the first humans to move from Africa into what is now the Middle East and then into Europe. They lived in these more Northern regions approximately 200,000 years ago. A great deal of what we know of them comes from the excavation of burial sites. The burial site itself is a distinguishing characteristic. Funeral remains provide clues to the religious beliefs and the social status of the person. How bodies were buried reflected their power and importance in a group. Ritual relics with skeletal remains showed the existence of a meaning system and perhaps belief in an afterlife.

For our interests, early human remains show that fellow humans were treated very differently from other animals. They show that these earliest human beings cared for their sick and deceased fellows. In some primitive cultures there is evidence to suggest that a type of drug or drink was often used to ease the passage from dying to death or perhaps even to bring about death. The Greek term *pharmakon*, from which our word pharmacy derives, originally meant *poison*. Primitive healers were better known for their use of poisons than for any curative remedies they had at their disposal. The term euthanasia, however, was not used to describe what these healers did with their patients. But it seems clear that the practice of taking life did exist and that healers were often the ones to do so.[110]

Poisons were used by persons in authority who usually combined both religious and medical powers with their political status. The idea of a community or political leader being actively involved in a person's death is not new. What is new today is that euthanasia and assisted suicide are being discussed and debated both in legislatures and in the public forum. Historically, however, the practices were not called euthanasia or suicide.

The terms euthanasia and suicide may be relatively new, but not so the practices of hastening death and causing it. The moral acceptability of these practices depended upon religious beliefs and community customs. Over the years, religious beliefs change and so do community customs. Beliefs evolve, cultural attitudes and customs evolve, and so do the meanings of terms like euthanasia and suicide. It makes sense to take a short look at this evolutionary process in preparation for looking at today's cultural debate about the legalization of these practices in the U.S. and in Europe.[111]

DEATH PRACTICES IN EARLY CULTURES

In some indigenous cultures, dying persons were exposed to the smoke of a smoldering fire which gradually caused a loss of consciousness and then death. In migrant or nomad tribes, old persons who could no longer travel or contribute to tribal survival, said good-bye and went off alone to die. Whether these practices were understood to help a person to die or actually to relieve the community of a burden is difficult to determine because this distinction is a more modern development. What is clear from the historical record is that nomadic tribes did not usually have invalids or chronically ill members. It appears that they applied the Darwinian principle and the law of the jungle: only the most fit survived.

In some Latin American indigenous cultures, there was a tribal pain reliever who did his job by actually breaking the backs of persons in pain.[112] In so doing, he relieved the pain and in the process took the patient's life. Once vertebrae were broken, the pain reliever laid the patient on his back, made the Sign of the Cross over his body and recited certain prayers. Then the body was delivered to the family.

In many primitive tribes and cultures, it was considered both humane and necessary to take the lives of some small children. When communities were forced to flee because of a threat, they often killed their small children. Plutarch tells us that the Spartan culture took the lives of deformed and sick children in order to maintain their military supremacy over other groups.

LATIN AND GREEK CULTURES

Although the terms euthanasia and suicide are Greek and Latin respectively in their roots, they are, as we have seen, relatively modern in their current meanings. The focus today is on the act of taking life and on the pressing question of whether such an act by self or others is ethical. In Antiquity, the focus was different. Euthanasia referred more to one's mental or spiritual state: whether the person was composed, self-controlled, psychologically balanced at the end. The term did not carry all the ethical connotations that it does today. In Pre-Christian Greek and Roman cultures, achieving this spiritual/psychological euthanasia may have involved some form of help. Euthanasia was a way of dying, not an act which brings about death, and not identified with killing. If death is natural, then aiding nature in the sense of helping a person to die a good death was not seen as morally wrong. The term euthanasia did not mean taking a life.

The support in Antiquity for what today we call suicide, i.e., administering of the means of death by oneself, was widespread. Even what we call active euthanasia, the administering of death by another, was generally permissible if the circumstances were right. Philosophical and religious support came from many different sources. Opposition to an active taking of life, however, was not completely absent in classical cultures. The Pythagoreans, for example, believed that earthly existence, including pain during life and at the end of life, was meaningful. Each life had a natural span which had to be respected. Early termination of life by followers of Pythagoras was considered wrong.[113]

Plato disagreed with the Pythagoreans. He used different arguments to justify his stand. In the *Republic*, he insisted that citizens be healthy and that those who chronically needed medical help simply should not be treated. The chronically ill were unable to contribute either to their own development or the state.[114] The doctor's role was to facilitate their removal by denying them any kind of medical help. Death was considered preferable to life when a person was unable to contribute to society. Doctors decided whether a person could contribute, and if the decision was negative, the doctor's role was to deny medical help.

Aristotle disagreed with Plato. He was opposed to taking life even when a person was suffering from an incurable illness.[115] He used the same background concept of the person as a social being, but for him it was duty to the state that made taking life immoral. Not only did taking life deprive the state of one of its own, but it was an act of cowardice for Aristotle. Dying courageously meant not giving in to death. Virtue was tested in suffering, and each citizen's courageous conduct was an example to others.

The stoics placed certain conditions and limits on the taking of life in order to make sure that it was not an impulsive act. The pain had to be great, and life with illness had to be incompatible with an individual's needs and self-understanding. The illness had to be incurable, and the person had to consider social responsibilities.[116]

Even in classical Greek and Roman societies in which the taking of life was common, the understanding was that the patient was capable of making a reasoned decision. Agreement was strong that a taking of life could not be an impulsive act. It had to be a reasoned decision. It could not be an unvirtuous act of cowardice nor a flight from social responsibility. There was the same strong agreement that dying should not be prolonged, and futile medical interventions had to be avoided under any circumstance. Nature was the objective foundation of morality. Nature had to take its course, and sometimes it moved toward death. Delaying this process was bad medical practice and bad morality. In the classic cultures of Greece and Rome, what today we mean by active and passive euthanasia were both widely practiced. This morality changed somewhat when Christianity gained converts and afterwards political power.

Certain practices from the pagan era, however, were approved in Christian culture because they met the standards of Natural Law morality. What we call passive euthanasia became mainline Christian moral practice. When treatment is futile it should be withheld. Even what we call active euthanasia was commonly practiced on soldiers wounded in battle. Following a battle, fields were frequently strewn with dying men for whom there was no medical help available. Often they lay dying and groaning in agony. These soldiers were often finished off by fellow soldiers out of compassion. Warriors in fact, carried small knives in their belts which they used in such circumstances, and the name of this weapon was misericordia, i.e., mercy. The act itself was called a "thrust of mercy," or an "act of grace."

One common thread in this short history of what today we call euthanasia and assisted suicide is the theme of pain relief. Given the absence of medical sophistication, the only way pain could be relieved in some cases was by taking a suffering person's life. Official perpetrators of this act did not ordinarily enjoy great respect or social status. For a long while these killing acts were done not by healers but by witches and sorcerers. Later on, physicians entered the scene and began a medicalization of the practices. Once this happened many things changed.

PHYSICIAN INVOLVEMENT WITH THE TAKING OF LIFE

The earliest ethical rules for doctors were so noble that obedience to them raised healing from a work to a profession: indeed, to the highest of the professions. One of the earliest summaries of ethical rules was the Hippocratic Oath.

The Hippocratic Oath reflects either the influence of Pythagorean religiosity or a reaction against widespread killing by physicians, or both: "I will neither give a deadly drug to anybody if asked for it, nor will I make a suggestion to this effect." This public promise represented commitment to a morality which dissented from common medical practice. Obviously, people did ask for death, and physicians sometimes suggested that a person's life be taken. No doubt, their poisons were more effective than their medications.

Medicine in the words of the oath was an *art*. Later on in a different Hippocratic text (Aphorisms) we read: "The art is long; life is short." In yet another text (Precepts) we read: "Where the love of man is, there is the love of the art." These Hippocratic texts help us to understand why this difficult art is inseparable from the highest morality. The doctor can do great benefit but also great harm. The rules and standards in the Oath reflect both possibilities. Proscriptions in the Code were against common practices by ordinary healers. The ethical standards of the Code became the core of humane western medicine.

Elevated and demanding ethical rules separated the Hippocratic physician from other Greek healers. The latter were known for their use of poisons. They assisted in suicide and actively took patient's lives. The Hippocratic physicians, on the other hand, distinguished themselves by promising to do neither.

After promising to use the art for the benefit of the sick person, the physician promised to avoid doing harm or injustice. In the many other Hippocratic texts, the physician promised to exercise special sensitivity in order to avoid harm while treating the sick. He promised to help (beneficence) and not to hurt or harm (non-maleficence). He promised not to administer a poison and not to give a deadly drug. So firm and strong was the commitment not to take life that it extended even to fetal life. The Oath has gone through changes from time to time but the commitment not to take life has not changed. It fit perfectly with Christian and Jewish morality, which contained a similar strong proscription against intentional taking of life.

The anti-killing stand, however, did not fit so nicely with the reasoning of respected Greek thinkers. The background metaphor used to understand society by philosophers like Plato was a physical body. Persons were compared to social body parts and expected to perform useful functions for the "body politic." Life without making a social contribution was considered not worth living. Chronic illness was thought to require too much care to be worth the trouble either for society or for the individual. Medical treatment was considered "wasted on a man who could not take care of his ordinary round of duties and was consequently useless to himself and to society."[117]

It is worth noting that even with his strongly social view of life and medicine, Plato over and over in *The Republic* argued for not treating when intervention was futile or unable to accomplish what he considered to be the ends of medicine (what later will be referred to as passive euthanasia). Only once, in a section of *The Republic* in which he was discussing education and what makes a good judge, did he pass over the line and endorse the active killing of certain patients. First he said, "The physically unsound they will leave to die."

Then, as an afterthought or in a throw away remark and completely out of context, he said, "They will actually put to death those who are incurably corrupt of mind."[118] He may have said these words in one line of an important book but he did not defend or even explain the idea that physicians should be judges of mental and moral corruption and should remedy vice by killing the morally corrupt. Frankly, I wonder if this line could have been added by one of his scribes.

Averroes,[119] the great medieval Muslim philosopher, wrote a commentary on The Republic. He mentioned of course Plato's metaphor of a body politic and the requirement of contributing to society. He followed Plato in recommending that treatment not be used for incurable illnesses. Rather, the incurable patient should be permitted to die. But he did not mention the line about killing certain patients. Both Plato and Averroes seemed to have understood and respected the distinction between not harming patients with futile treatments or not extending the dying process when it is burdensome (passive euthanasia), and actively taking a dying patient's life. This distinction was certainly respected in Christian culture.

JUDEO-CHRISTIAN CULTURE

Greek culture and moral philosophy were focused on the cultivation of virtue and the importance of reason for understanding reality and for determining right and wrong on the basis of that understanding. The Judeo-Christian culture continued this tradition, but the reality of dying took on a completely different meaning. Consequently the morality of dying also changed. Life in a Christian perspective was a gift of God, a creation in the image of God, an object of divine providence. Even suffering and dying were believed to be ordained by God. Both the time of death and the type of someone's death were divinely ordered. Suffering, rather than being something unnatural and disordered, was an imitation of Jesus, an instrument of salvation, and an exercise of virtue. Christians were called upon both to relieve suffering as Jesus had done so often during his ministry and sometime to bear suffering. Dying was an antechamber to resurrection and eternal happiness. Different understandings of the end-of-life reality changed the morality of death and dying.

The Jesus whom we meet in the Gospel was not a moralistic person. He constantly drew criticism from moralistic and righteous members in his Jewish society. But Jesus did provide moral norms, moral ideals, and a living example of moral goodness. For example, he reminded his listeners of the commandment not to kill, and then went further. "You have heard the commandment imposed on your forefathers 'You shalt not commit murder; every murderer shall be liable to Judgment. What I say to you is: everyone who grows angry with his brother shall be liable to judgment: any man who uses abusive language toward his brother…etc.," Matthew V-21, 26. Jesus' moral standards and examples influenced the moral standards of Western society. Taking a life was understood as usurping a divine right. This reinforced ethic against killing made killing even at the end of life morally wrong.

When Constantine made Christianity the official religion of the Roman Empire in the 4th century, the strong pro-life ethic of the Judeo Christian moral tradition became the public morality. The Hippocratic norms against taking a patient's life had already gained the respect of early Christian writers (like Jerome), but after Constantine, the Hippocratic tradition blended with the Judeo Christian tradition to form the empire's official civic morality and medical ethics. By the Middle Ages (12th Century) the high Hippocratic standards which had been distinctly minoritarian in pre-Christian Greek culture, became mainline medical ethics throughout Christendom.

We can see examples of the Judeo-Christian ethic as early as the 2nd century A.D. An important early Christian document, *The Shepherd of Hermes* (circa 150 A.D.) urged Christians to care for poor people at the end of life lest they commit suicide. Later on, St. Augustine, perhaps the most influential theologian of Christendom, took a stand against suicide based on the commandment "thou shalt not kill." He also argued that suffering must be endured and that the time and circumstance of death is in the hands of God. Because of Augustine's dominant role in theology, priests and confessors throughout Christendom taught that self-killing as well as the taking of a dying patient's life were wrong. Augustine's influence is hard for us to imagine today. Sad to say, however, he did admit an exception to this proscription based on the right of political authority to wage war. His teachings also help to explain the scandals of the Crusades, and the Inquisition, and finally the Holocaust.[120]

The other giant of Christian theology was Thomas Aquinas. He added other arguments to Augustine's basic point of God's domination over all human life. For Thomas, suicide was also wrong because it contradicted a built-in structure and basic motivation to preserve life. To this he added Aristotle's idea that it violated an individual's duty to society.[121] Suicide (and what we understand now as euthanasia) violated God's love, the love of self and the duty to community. In Thomas' moral reasoning the commandment "Thou shall not kill" coincided with and was supported by arguments based on reason's recognition of natural self-love and commitment to community. Killing oneself or another was considered immoral based both on divine law and on Natural Law.[122]

A good death in Thomas' time was understood to be a tranquil and accepting death, a resigned and peaceful death. A good death in this sense was achieved by support, pain relief, caring acts and prayer, but not by actually taking the person's life. The physician became less and less important at the deathbed. Clergy, family and friends were the instruments of this understanding of euthanasia.

Christians were obligated to relieve suffering, and the involvement of religious men and women in hospital work attested to the importance of this obligation. In imitation of Jesus, they committed themselves to relieving suffering and especially to caring for the dying. Care and pain relief, however, did not extend to embracing the Greco-Roman practice of using inflicted death as a way to relieve pain.

The dying process seems always to have generated ethical concerns about the right way to help. Influential thinkers like Sir Francis Bacon in the 16th century urged physicians to do more to relieve a dying patient's pain, even when it "may serve a fair and easy passage."[123] Francis Bacon re-introduced the term euthanasia into English to refer to a dying with adequate pain relief provided by physicians, but he left out of the meaning active killing, even mercy killing.

In the same century, Sir Thomas More in his Utopia,[124] reintroduced elements of what was meant by a good death in the classical pagan period. In More's ideal society, he spoke of patients "freeing" themselves from painful existence or permitting others to "free" them. The term euthanasia was not used in English, but obviously he was speaking of what euthanasia had meant in pre-Christian Greek and Roman culture. Still there was no mention of a physician's role. And concerns were expressed about the possibility of abuse.

Later on in the 16th century, Protestant Reformers reinforced the Augustinian viewpoint on the morality of death and dying. They insisted that directly taking life either by self or others was the antithesis of faith, and without faith there was no salvation. The reformers looked to biblical texts for arguments to oppose a revival of Greek and Roman cultural practices during the Renaissance. The focus of Protestant theology was on sin, original sin and the fall. Illness and suffering in this perspective were understood as punishment for sin. The challenge of dying was understood to be endurance of suffering in good faith. A tranquil and easy death came to look like a temptation to be avoided.

The Renaissance humanism of the Catholic world resurrected a milder version of what a good death had meant in the classical period. The focus was not on sin and suffering as punishment for sin, but rather on a good death as a more natural death. Again there was a role for the physician: to relieve the pain caused by the "unnatural" state of disease. The physician recovered his role and his place at the bedside of a dying patient. He was to make possible a good death in the sense of a tranquil death, free of pain. Science and medicine were to control the body functions so as to make possible a painless end of life. What the doctor should do and could do for a dying patient re-emerged in Catholic moral theology.

Physician involvement in euthanasia referred usually to withdrawal or stopping of treatment once a patient's illness was understood to be beyond medical remedy. In the pre-Christian pagan period, physician involvement may have meant the active taking of a patient's life. During the rebirth of this period in Christian history, however, it did not mean active killing, but rather helping a patient to die peacefully by withholding or withdrawing remedies and by providing pain relief. Doctors were expected to practice euthanasia in this sense.

Catholic moral theologians in the 16th century distinguished between "*ordinary*" (effective and not burdensome) and "*extraordinary*" (futile or burdensome) treatment as a way of justifying letting persons die peacefully.[125] Patients could refuse extraordinary treatments, but not ordinary ones. Later, they also justified using adequate pain medication, even when adequate pain medication might repress respiration or send the patient into a deep sleep ending in death. Pain relief was considered the right, proper and moral thing to do. A good death may very well require withholding or stopping medical treatments and as much pain relief as necessary to die in peace. Catholic moral theology, however, drew the line at direct, intentional killing. This ethical tradition was reflected in civil law which drew the same lines. Until most recently all states in the U.S. had laws against active killing, even of dying patients, and against assisting them to kill themselves. We see the same influence of Judeo Christian ethics on laws in other cultures and nations.[126]

Christian moral teaching about death and dying began with the Judeo-Christian commandment not to kill and over the centuries added subtleties to address new problems.[127] The negative obligation (do not kill) was weakened by an exception for war but strengthened by an exception for defending against unjust attack on innocent life. "Do not kill" was never understood to mean that a person had to do everything possible to avoid dying or to preserve life. Without violating either the 7th Commandment or Jesus' added standards (Matthew 5: 21, 26), a person may withhold, withdraw, refuse, forego, any procedure or intervention which is 1) not likely to be successful, 2) difficult to accept, 3) frightful, 4) painful, 5) costly, 6) repugnant. These distinctions and subtleties developed over the course of centuries by moral theologians came to be respected by religious and secular persons in both the private and the public realm.[128] They became an important part of Western medical ethics.[129]

SECULAR ENLIGHTENMENT CULTURE

The 18th century brought a revival of pre-Christian pagan culture: a type of second Renaissance. The Enlightenment meant faith in reason, science, human progress, rather than in the doctrines of revealed religion. The anticipated progress was expected in great part from abandoning what Enlightenment thinkers described as the ignorance and superstition of religion. During the 18th and 19th centuries, little by little, the scientifically trained physician replaced the priest in power and prestige. His expert help was sought in order to ward off premature death as well as to relieve pain. He took the place of the priest/minister/rabbi at the bedside of dying persons.

The focus of modern medicine is on illness rather than on the person of the patient (the opposite of what we described as a more humane medicine). Gradually criticisms began, even as progress and power accelerated. Once there was nothing more a scientific physician could do to offset the inevitability of death, he simply withdrew, leaving the patient to die alone. Critics called attention to this. They urged doctors to be more humane, to avoid continued use of aggressive treatment to the end, to provide comfort measures, and not to abandon the dying patient. These critical voices, however, were drowned out by an enormous explosion of medical science and new technologies which continued to be applied even during the dying process. During the 20th century, patients ordinarily died in a medical setting, surrounded by medical technology, often in pain, always in isolation: i.e., a bad, indeed a worst possible death. These inhumane practices (abuses in many instances) fueled a new movement for euthanasia, a good death. In the latter half of the 20th century, euthanasia was back as a major issue in medical ethics and a major topic of political debate. Now, however, euthanasia was understood to mean actively taking a patient's life.

Drugs to relieve pain had been known for centuries but were not commonly used by physicians, even for patients in pain at the end of life. Morphine was isolated in 1816 by Fredrich Wilhelm Serturner, and the invention of the hypodermic syringe made possible its rapid absorption and quick action. Safe and effective control of pain has been possible since the mid 1800s.[130] And yet, the post-Enlightenment scientific physician was more interested in learning from symptoms as indicators of disease than in the elimination of symptoms through palliation. Government vigilance of physician prescriptions of drugs also had the effect of restraining the use of pain medications. As more and more individuals and families had experiences with patients dying in pain, active euthanasia became both a medical ethics problem and a political issue.

In the U.S. one of the first explicit endorsements of the pre-Christian practice of taking the life of dying patients came in an article by Samuel Williams entitled "Euthanasia"[131] in 1870. The conditions and precautions which he expressed are the same ones we hear today. The killing must be consented to (willed), the patient must be dying and in pain. For Williams taking the life of a patient in these conditions was a duty rather than a violation of duty as the Hippocratic tradition had insisted. In his article, no mention was made of the presuppositions which had informed the anti-euthanasia ethic of Christianity:

i.e., life belongs to God, commandment against killing, pain nobly borne as virtuous, suffering as identification with Jesus and anticipatory of eternal reward. Williams was writing from secular Enlightenment beliefs rather than from Christian beliefs. The 20th century would witness an open clash between a secular Enlightenment vision and the older Judeo Christian moral perspective.

During the 20th century in Western culture, medicine became the primary profession and the doctor the most powerful professional. Scientific medicine continued to generate knowledge about illness and new ways of trying to win the war against disease. In many European nations, the Enlightenment secular perspective increasingly replaced the religious vision. In the U.S., religion undergoes periodic revivals, and still it is unclear whether the secular or religious perspective will prevail. The clash of perspectives plays out most publicly over moral issues like abortion and euthanasia. Following a recent Supreme Court decision, each state will have to develop its own legislation on euthanasia and assisted suicide. This means that the issues will be publicly debated throughout this century, if not the next millennium.[132]

Physicians, like other citizens, are divided on the issue, with a slight majority favoring legalization of active euthanasia in order to protect themselves against malpractice suits. Theology and theological arguments re-enter the debate and are discussed in newspapers, TV shows, university classrooms, etc. The criticisms of mainline medicine's record on caring for dying patients which first made an appearance in the middle 1800s, now permeate the debate and are recognized by both sides. Because there is no way to effectively mediate these divisions in society on the ethics of death and dying, the questions are repeatedly referred to courts and legislatures. Victory of one side or the other will depend upon which one can make a convincing case for remedying the problems generated by too many patients dying in pain or dying a bad death. At this time the AMA has begun a nationwide program to remedy some of the widely recognized problems in treating dying patients.[133] American Medical Association leadership has taken a stand against euthanasia by physicians and physician-assisted suicide.

MODERN MEDICAL INVOLVEMENT IN EUTHANASIA

Doctors became more intensely involved in the euthanasia debate in the late 19th and early 20th centuries. We already mentioned the fact that historically the term Euthanasia has had many different meanings and only relatively recently was used in English to refer to the active taking of the life of a dying patient. Before that, it referred to what we could call a psychological, or spiritual, or religious tranquility at the time of death. No doubt physicians actually administered death in classical Greek and Roman culture. But Christian moral teachings changed this practice for many centuries. In the late 19th and early 20th centuries, however, the idea of physician-administered death again began to be debated.

Physicians had to be able to make the diagnosis of incurable disease and then to predict accurately when a patient would die before it would be reasonable to suggest that a doctor take the life of a dying patient.

The idea of doctor involvement was promoted by the Williams article, and then by another essay entitled, "The New Cure for Incurables," by Lionel A. Tollemache. This second article directly confronted religious objections to doctor-administered death with secular reasoning, especially Darwinism. Tollemache talked about "an overcrowded population," a "struggle for existence," and the idea that chronically sick people are useless and crowd out healthier, more useful persons.[134] The idea of the sick as useless or strains on society and of doctors as solving this "problem" by taking their lives would be repeated in other European countries.

Even before the emergence of Fascism, books were published in Germany claiming that the state had a right to take the lives of the terminally ill, the mentally ill, the retarded, and deformed children.[135] The authors of one of these books (a psychiatrist and a philosopher) talked about killing as "hygienic," and as a "healing therapy." This same metaphor was used in a book entitled *The Right To Death* by Adolf Jost. His argument was that the state had a right to kill in order to keep itself strong and healthy.[136]

These ideas were widely held in Germany especially among physicians and more especially among psychiatrists. The Judeo-Christian moral tradition and the long Hippocratic tradition were both set aside or subordinated to the sciences in the new Enlightenment culture. Doctors killed an estimated 200,000 persons during the Nazi period, in part because a climate favorable to state killing preceded the Fascist state and made it quite easy for Nazi leaders to implement their eugenic policies of mass euthanasia.

The involvement of physicians with suicide and euthanasia is understandable despite a strong medical ethical tradition which had proscribed both. Doctors are more likely to witness the most painful deaths. And when real analgesics like morphine became available, there was even more reason for doctors to be called to attend the dying. As soon as ether was used successfully for surgery, doctors speculated about its use for "mitigating the agonies of death." Morphine was extensively used in the U.S. Civil War, and medical journals discussed the use of ether, chloroform and morphine to manage pain for dying patients (i.e., *British Medical Journal* in 1866).

Once pain relief for the dying became medically possible, it was a short step to using the same medications actively to take a patient's life. Proposals to use analgesics in this way re-enkindled the euthanasia debate which had been held in check by the power of Christian moral teachings. These were now weakening because of the spread of Enlightenment secular views and the use of Darwin's evolutionary theory to support eugenic projects. Because doctors were the only professionals authorized to use these drugs, the AMA got involved in the public debate.[137] They took a strong stand against the proposals which they claimed would turn physicians into executioners, a position they continue to defend today.

The arguments on both sides of the debate were not much different then from what they are now. Also noticeable was a certain resentment against the power which physicians exercise and concern about their using it to extend dying against a patient's wishes. A bill to legalize euthanasia was defeated in the Ohio legislature in 1905-1906, by a 79 to 23 vote. After that, the issue pretty much dropped from public attention.

By the 1930s the Euthanasia Society of America was operating in New York and sponsored another bill to legalize euthanasia, this time "with adequate safeguards." This bill too was defeated, but again the public debate was rekindled. Then, as now, the basic driving force for legalization was pain. If pain at the end of life is adequately addressed, other arguments like the patient's need on the part of patients for control are not likely to be successful. Physicians were practitioners of euthanasia in the past and some have been public advocates today. Professional associations of physicians in the U.S., however, have always recognized the negative consequences for the profession if doctors become killers and have mounted campaigns to defeat legalization efforts.

Negative reaction to Nazi euthanasia programs quieted the debate for a decade or so but by the 1960s the issue was back in the public forum. After the practice was first decriminalized (and then formally legalized) in the Netherlands, the legalization possibility became more realistic. The bizarre public behavior of Dr. Jack Kevorkian also helped to rekindle the public debate and force the issue of legalization.

The debate within the medical culture exploded after a small, one-page account of a young resident who took the life of a dying patient whom he didn't know, after being awakened in the middle of the night.[138] The short clinical case was not signed. The young resident's identity was not revealed. A lethal dose of morphine was used to take the patient's life. The idea that the official journal of the AMA would publish such a piece was a mystery to critics. Following publication of this short piece, The Journal of the American Medical Association (JAMA) was on the front page of papers around the country. The press attention was such that not long after this, the New England Journal of Medicine (NEJOM) came out with its own version of a euthanasia endorsement from within the culture of medicine. Dr. Timothy Quill told of assisting one of his patients to commit suicide.[139]

By the 1990s public debate in the U.S. had become intense. There were Kevorkian-generated court cases, legal initiatives in Michigan and Oregon, decisions against anti-suicide laws by federal judges, etc. These latter were appealed finally to the Supreme Court which decided that there is no constitutional right to suicide or euthanasia. If these practices are to be legalized, it must be done at the state level. This means that the issues and arguments and distinctions which have been debated for years will now be presented to ordinary voters for their judgment: (1) whether there is a distinction between the power to choose and the right to freedom; (2) whether the right to freedom supposes a background vision of good life which freedom promotes; (3) whether individual rights to autonomy and privacy apply to suicide and euthanasia; (4) the distinctions between actively killing and respecting refusal of interventions to let patients die; (5) the use of pain relief even when the dosage may repress respiration; whether legalized suicide and euthanasia will create a culture of death.

In the U.S. most states claim the right to take certain lives (capital punishment). It is, however, mainly individuals in the new debate who claim to have a right either to be killed directly by a doctor or to receive medical assistance to kill themselves. Euthanasia as it was practiced in pagan Greece and Rome is back. We saw in classical Greek and Roman culture, and again in 20th century totalitarian regimes, justification for the state's right to kill. In the U.S. today it is a different kind of right or power that is being claimed. Now it is the individual whose power and right reaches the point of self-killing with doctor assistance, and the point of demanding that doctors actually kill certain patients.

Euthanasia under one or another understanding of the term has been with us forever, and as we begin the third millennium in the U.S. and Europe it is very much present to public consciousness. One key point is whether the claim of an individual right to kill can remain individual or will inevitably become a state right. Another key point is whether the distinction between active and passive euthanasia which has been part of Judeo-Christian and Hippocratic ethics can hold. If not, what we understand as humane medical care at the end of life and a traditional Christian moral teaching about what is ethically appropriate at the end of life may both become marginalized instead of mainstream.

JUDEO CHRISTIAN BELIEFS AND PUBLIC MORALITY

For persons who continue to follow either a Judeo-Christian or a Hippocratic medical ethics, the case for not killing other human beings is fairly easy to make. The foundation of Judeo-Christian morality is a covenant with God and an accompanying moral code called the Decalogue or the Ten Commandments. Commandment number 7 is clear: "Thou shall not kill." The Ten Commandments lay out a kind of bill of rights and obligations without which no society or political body can stay together. People will not come together into community without a strong guarantee that they will not be killed. Some version of the seventh commandment we find in every society.

The Mosaic Law against killing was weakened during the period of the judges (1200s, or 1100s BC) and later under the kings (1000 BC) because of exceptions related to war and justifications of brutal treatment of enemies. Over the centuries, however, many of the exceptions to the proscription against killing have been eliminated. Morality, like everything else, evolves; and the evolution has been toward the establishment of a formal, international right to life that proscribes any direct and intentional killing. A similar evolution took place in theological ethics which also gradually removed earlier justifications for killing.

Catholic Natural Law ethics relies both on critical reasoning and Biblical exegesis. Suicide and euthanasia have no biblical foundations. Early in Genesis we find the story of Cain and Abel along with other Genesis stories which ground the view that human life is a gift of God and which condemns the intentional taking of another person's life. Even the story of Job, the paradigm of a person in unbearable pain, gave no support to relieving distress by taking one's own life or asking to be killed. Rather, Job gave meaning to his suffering, a meaning founded on faith and hope.

The teachings of Jesus reinforced an evolved biblical proscription against killing. His Good Samaritan relieved the pain of a dying man by providing relief, not by assisting in suicide or practicing mercy killings. Jesus Himself was pursued throughout his life by enemies who wanted Him killed, but He never responded by justifying the killing of his enemies. Never did He call for anyone to be killed. Nowhere did He support either suicide or euthanasia. The anti-killing ethic of Jesus easily integrated with the same ethical stand of the Hippocratic physicians to create a 2500-year-old anti-euthanasia and suicide tradition in Western medical ethics.

What we know of Jesus from the gospels we find reflected in early Christian writings, especially the Pauline epistles. In his letter to the Corinthians, Paul took a stand against both Greek and Roman cultural perspectives, in his way of understanding community and its members. Whereas Plato had given little worth to individual members of a society, Paul insisted that every member of the Christian community (the Body of Christ) was precious. Paul and Plato both used the body analogy to understand community, but the way this analogy played out in each thinker was very different. Plato endorsed eliminating body parts (persons) which did not meet his standards of health and productivity. Paul saw an important role and contribution for every member. "Even those members of the body which seem less important are in fact indispensable." "We honor the members we consider less honorable by clothing them with greater care." "God has so constructed the body as to give greater honor to the lowly members...." "If one member suffers, all the members suffer with it…." (Corinthians, XII, 22-26).

The strong insistence in Judeo-Christianity on respect for life stood in strong contrast to common attitudes in pagan cultures. In Greek and Roman cultures this life was all there was. The role of the gods was to help persons realize their hopes and dreams for this life. But human life did not have a sacred value. The practices we see in Roman and Greek cultures show just the opposite.

In ancient Rome, killing was a sport; in fact, it was the sport. Killing for ancient Romans was what football is for modern persons. When the emperor finished the Roman Coliseum around 70 A.D., the opening events were all about killing. Gladiators killed one another. Vicious animals killed one another and then killed human beings. Later on, teams of warriors in chariots fought one another to bloody death, and fans in the amphitheaters lined up for and against the different teams. Riots sometimes broke out and then the fans of one team killed those of their rivals. Roman culture lasted many centuries and was characterized by a lust for blood, violence and public killing of human beings. Roman law even permitted the killing of women for certain offenses against men, like infidelity.

This moral culture began to change when Constantine made Christianity the official state religion. But Constantine himself did not become a Christian until just before death. Presumably he had seen visions and had benefited from divine interventions, but he had never personally adopted the Christian faith. One explanation for this anomaly is that Christianity meant not killing other human beings and Constantine thought he needed to kill in order to rule.

Before Constantine (4th century), emperors killed their enemies and were themselves killed. During the 1st century A.D. almost every emperor was killed either by himself or others. A certain consistency is evident in the practices of killing in war, in politics, and in the sports arena. Soldiers killed, politicians killed, athletes killed, and people lined up in favor of one or another team and killed one another. In one particularly bloody event in the Coliseum in 107 A.D., 10,000 human beings and 11,000 animals were killed.

The stand of Western medical ethics against killings cannot be appreciated without taking into consideration the background beliefs of the Roman culture which preceded it. The contemporary debate about whether or not to reject the changes in public morality which were brought about by Judeo Christianity cannot be intelligibly conducted without taking history into account. The change from killing for entertainment to prohibition against killing, even state killing of criminals, was a gradual evolution and an example of moral progress. If this is so, the question now is whether we should reverse course.

Killing has for many centuries been against the common law and statutory civil law. Why, then, is there such a successful effort under way to change what has for so long been a crime into a right? The answer, I believe, is because of changes in medicine: people have seen or heard about patients dying in pain, in isolation, and tied to intrusive medical technologies.

Contemporary Protestant, Catholic, and Jewish morality all agree that direct taking of human life is wrong, but this religious solidarity is no match in contemporary secular culture for the fear of losing self control and being forced into a protracted and painful medical dying. People are afraid. Unless that fear can be relieved, laws will be changed, which means the end of a humane public and medical ethics, and a return to publicly approved killing.

PHYSICIAN-ASSISTED SUICIDE

Because the objective of modern medicine is to conquer death, the overall well being of the patient is often lost in aggressive fixations by specialists on eliminating disease from separate parts (organs, tissues, cells, genes). The narrowed focus, a commitment to preservation of life, and the availability of ever more powerful technologies explain why so many people today have extended life spans. These conditions also explain why the same people all too often have extended dying processes. Many patients die terrible deaths, tied to technologies, isolated in hospitals, all too often in pain. The narrow fixation on physical parts rather than on the whole person's quality of life, plus ignoring the fact that death is inevitable, explains why even permanently comatose patients are often kept alive for weeks and months and years. People who see such things are nearly unanimous in their reaction: "I don't want to die like that." Modern medicine's technological triumphs have demanded a great price: the medicalized dehumanization of the end of life. What people fear most and what is most objectionable ethically is an extended and painful dying process. Fear of this dehumanization leads many today to demand the right to be assisted in committing suicide.

More than two million people die every year in the U.S. Over 80% die in a health-care facility. Death occurs following a decision either to withdraw a technology (to stop a procedure) or to withhold a technology (not to start a procedure). Death and dying have become increasingly a matter of medical decision-making. And yet only recently have efforts been undertaken by the AMA and other professional associations to train doctors to make decisions at the end. All too often in the past, once their powerful technologies ceased to work on the patient, and the family refused the next technology, doctors left the scene. The typical response was "there is nothing more I can do." Now doctors are being trained to manage death by managing palliation and comfort measures. They are encouraged to stay with the dying patients. They are taught when stopping treatment is appropriate in order to let the patient die.[140] Some doctors, however, refuse to change their practice of aggressive treatment to the end. Others are willing actually to take the patient's life or to assist in a patient's suicide.

Almost inevitably patients have to be helped to die. The ethical question is: how far should the help go? Does the help extend to physician-assisted suicide? What are the ethical limits to what doctors can do and yet remain true to personal moral beliefs and professional codes? Whether medical professionals like it or not, they have to develop some ethical sophistication and learn to respond appropriately to issues of death and dying. In particular, they have to think through the question of whether or not to assist in suicide and to perform euthanasia.

The meanings and moral directives provided by religion have largely been left aside today in favor of scientific belief systems and nonreligious foundation of morality. According to secular beliefs, this life is all there is. Secular beliefs affect the attitudes of people toward living and dying just as religious beliefs do. If aging and sickness are burdensome, costly, meaningless, and if this life is all there is, then getting out of the way by suicide is reasonable and moral. If this life is all there is, some will want to hang on as long as possible. Others, however, will prefer to adopt a traditional pagan approach: after weighing the benefits and burdens, a decision is made either to continue living or to commit suicide with help from a physician.

Today in the U.S., physician-assisted suicide is being proposed as the way for elderly persons to gain the control which they need over dying. It is a common belief that if life is miserable, one has the right to get out, with a physician's help if necessary. The burden of proof for a right to suicide has shifted to those who oppose it. Stories abound about how some sick and elderly people are kept alive at the end by unwanted medical interventions and are assaulted against their will by intrusive procedures. The fact that such abuse occurs is enough to push many elderly persons to choose suicide with medical assistance and to support the campaign being waged to make physician-assisted suicide a solution to the problems with aging and dying.

To experience death even from a distance changes us. To see the limits of life in one's own aging or even in the aging of loved ones, forces one to ask about the meaning of life generally, and the meaning of one's own life in particular. To be in touch with aging and dying means to feel one's own history in a new way. It reveals the frailty of life and creates awe in the face of life's mysteries. But if aging and death are covered up as they are in our society, this too affects the life experience. It creates superficiality, artificiality, a hollowness in life. Would it make sense in today's society to make suicide with assistance from a doctor a right?

In the U.S. physician-assisted suicide is sharply distinguished from physician-administered euthanasia. To date Oregon, Washington State and Montanta, are the only state to legalize physician-assisted suicide, but other states are considering similar legislation.[141] Changes are taking place under the assumption that suicide is an individual act which is a quick and cost-effective solution to many problems of aging and dying. Judgments about the morality of suicide are made on the basis of a modern, secular, ethical standard: i.e., right is what I freely choose. The issue of suicide, however, is much more complex than a matter of personal choice, and it makes sense to look more carefully at the idea that legalized suicide is a solution to problems of dying. In a Natural Law perspective, this more careful look begins with a historical understanding of the reality.[142]

SUICIDE IN HISTORY

Many animals engage in risky behavior, but it appears that human beings are the only animals that commit suicide. Not only do human beings kill themselves, but they seem to have done so in every cultural context and every historical period. Fifty-four times, in 37 of Shakespeare's plays, a character commits suicide. "To be or not to be, that is the question." And Shakespeare was writing in a Christian culture which was strongly opposed to this practice.

In pre-Christian times, different cultures had different valuations of suicide, but frequently it was considered acceptable. Eskimo and other indigenous American cultures encouraged suicide and considered it to be a virtuous act. Nomadic societies valued suicide, especially by aging members, because it promoted tribal mobility. Socrates drank hemlock rather than compromise his integrity. Suicide by military men rather than surrender to an enemy was honored in ancient Greece and Rome and in most other cultures. In Palestine, during the siege of Masada, over 900 religious Jews committed suicide rather than surrender to Roman forces.

But not all pre-Christian societies valued suicide, and not all pagan philosophers supported it. Aristotle thought that suicide was cowardly and a violation of a citizen's social obligation. Roman law made it difficult to pass on property after suicide. It was, however, primarily the influence of Christianity on Roman Culture which made suicide illegal.

Life in the Christian perspective is a gift of God and not something over which individuals have ultimate decision-making power. Human life is made in the image of God. Suicide is considered a violation of the symbol of God's image as well as the commandment not to kill. Widely-held religious beliefs have an influence on law in every culture, and Judeo-Christian moral teachings about suicide certainly influenced the anti-suicide legislation which has been pervasive throughout the Western World.[143]

St. Augustine strongly opposed suicide. His writings had a major influence on civil law in the Roman Empire and later on the Protestant reformers. St. Thomas Aquinas added several arguments to those cited by Augustine. Suicide, he thought, is an act contrary to the natural inclination to preserve life, and therefore it was contrary to the Natural Law. Aquinas argued that suicide usurped the function of God and deprived society of important activities.

Protestants did not dissent from this particular Catholic teaching. Influenced by John Calvin, Puritans considered suicide a sin and separated the bodies of suicides from community burial grounds. Luther condemned suicide and so did John Wesley. Denials of religious burial in Catholic law continued in mainline Protestant practice. The strong stand taken against suicide by Catholic and Protestant theologians had a big influence on civil law both in Catholic and Protestant cultures.

Gradually, however, the secular legal systems adopted a more moderate stance. Suicide was not legalized in the West, but neither was it criminalized.[116] It was considered more the result of a mental aberration or of an emotional pathology than a wicked act. By the 18th and 19th century, most European countries had decriminalized suicide. The last to do so were England and Wales in 1961 and Ireland in 1993.

SUICIDE AND LAW IN AMERICAN CULTURE

In the U.S. today, a campaign is under way to leap from decriminalizing suicide to making physician-assisted suicide a basic human right. Decriminalizing an act, however, does not automatically turn the act into a right. A person may want to kill himself and be able to do so because it is no longer a crime, but this does not mean that he/she has the right to do so. Two appeals-level Federal Court decisions, however, struck down centuries old laws against suicide. In *Washington v. Gluckberg*, on the West Coast, anti-suicide laws were judged to violate liberty and privacy. In *Quill v. Vacco*, on the East Coast, the anti-suicide laws were thought to violate equality. In both jurisdictions, federal judges decided that individuals should have the right to die the way they want, including the right to die by suicide. Proponents of the change argued that by declaring suicide to be a right, society would extend the autonomy and control of sick and elderly patients by reducing the power of doctors to intrude into their lives with unwanted interventions.[144]

The Appellate Court decisions were appealed to the U.S. Supreme Court, which reversed both lower court decisions.[145] The Supreme Court justices upheld the traditional distinction between withdrawing or withholding burdensome medical interventions and actively taking one's own life. There is no right to suicide in the U.S. Constitution, the Supreme Court justices declared. Their decision left the older anti-suicide laws in place. But the Supreme Court decision hardly settled the controversy.

The same Supreme Court judges recognized that state legislators could legalize suicide. Crusaders for a right to suicide began immediately to lobby in order to change state laws. Those against the change in the state law used the power of the Federal Government. In Oregon, the only state in which assisted suicide is legal, the Attorney General tried to invalidate the state law by claiming that physician-assisted suicide violated federal drug regulations. Their attempt failed. The Oregon state law remains in force.

The campaign for legalization of physician-assisted suicide in each state is carried out usually without using the word suicide, which for most people still has a negative flavor. It is carried out as a campaign for the "right to die." Suicide, in effect, is merged with a long established legal and ethical right to refuse treatment even when death may inevitably follow. By merging an act of self-killing with an omission of burdensome interventions, the crusaders throw out a classical legal and ethical distinction between acts and omissions. By doing so they merge an act of suicide with a long established right to omit unwanted treatments. Suicide, however, goes beyond the right to refuse burdensome treatment.

The "right to suicide" as an individual liberty is argued against doctors, nurses, hospitals who have objections against performing this act. People are afraid of death. They are also afraid of a prolonged dying process, dependency, senility, becoming a burden, loss of control. All these fears explain the interest in suicide and the talk of an individual right to suicide. The claim is made that medical professionals have an obligation to help with the exercise of this right. Should medical professionals be obliged to help patients commit suicide even when such assistance violates centuries-old moral codes and deeply held personal convictions? Is suicide such a basic right that it overrides the rights of other people? The argument in favor of a right to physician-assisted suicide in final analysis seems based on an objection against nature itself. If nature or fate or evil causes disease and disability then the medical profession has an obligation to remedy the situation by making ill-fated persons dead.

The arguments against physician-assisted suicide center around many different concerns.[146] One concern is about extension of the practice. Once a right is declared to something, that something expands. Even when legislation places limits on the right, the limits tend to give way to social and individual pressures. First, only those who are dying are said to have the right, but immediately persons not considered dying but nevertheless suffering and unhappy also demand the right. First, only those who are competent and express an uncoerced decision to die have the right, but immediately persons who are incompetent, retarded, senile and even unconscious are provided the service at the request of family members who may be exerting pressure on the patient. First, only those who have uncontrollable physical pain have the right, but immediately persons with mental and emotional suffering demand the same right. The inevitable extension of services claimed as a right is the reason why special attention to social consequences is required. If every American has a constitutional right to die, it will be difficult to justify any state-imposed limits: physical pain or terminal illness or competency to consent. Convincing arguments can be made that limiting conditions built into a law do not hold.[147]

Any extensions of a legalized right to suicide will have an effect on the whole society. The effect of such legislation, however, even on the elderly and the dying, is not adequately appreciated. Once the suicide right is declared and the practice expanded, elderly persons who need help with day-to-day living will feel that they are a burden and consequently will feel considerable pressure to request suicide. A legally approved and culturally expanded practice of suicide will make it difficult for sick and elderly people to justify continued living. The same pressure will be felt by the poor people, the uninsured, and cultural minorities (especially women). They will feel that they have an obligation to get out of the way. A positive right designed to provide the elderly with more control and more freedom will have the consequences of reducing control and eroding the freedom to live for many.

Elderly sick persons will first be coerced into believing that they ought not continue living. Children of parents who are dying slowly, HMO administrators trying to save on medical costs, doctors impatient with chronic illnesses: all these will put pressure on the sick and elderly to get out of the way via suicide. What healthy family members and doctors believe about a sick or elderly person's quality of life can easily come to replace the person's own evaluation of his or her life. The last patient whom Jack Kevorkian assisted (and taped the killing for television) may have freely asked to be killed, or he may have been calling out in desperation for help. Dr. Kevorkian's evaluation was that the patient's quality of life was low and that he would be better off being killed.

Legalized suicide can be expected to alter any society. With legislation, the state, in effect, will turn over to doctors and families and patients the use of lethal power. The use of lethal power by the state is already for many Europeans dehumanizing and a scandal. The U.S. population, however, still favors capital punishment. But what if the power to take life is extended to every person, family, doctor? What kind of society would that create?

SUICIDE AND LAW IN DUTCH SOCIETY

In November 2000, the Lower House of Parliament in Holland approved both physician-assisted suicide and voluntary active euthanasia by a margin of 104-40. In Europe the two practices tend to be joined. In April 2001 the Upper House voted to approve the legislation. In 2002, the bill became law. This legalized suicide and euthanasia legislation enjoys wide support both in the Dutch public and in the Dutch medical profession. Only about 10% of either group is adamantly opposed. More than 50% of Dutch doctors have already assisted in suicides and have actively taken the lives of thousands of patients. A Netherlands Voluntary Euthanasia Association has more than 100,000 members. Opponents of the legislation have taken to the streets in demonstrations but stand little chance of changing the law.[148]

The current suicide and euthanasia bill applies only to physicians. The guidelines built into the bill are the work of the Dutch Medical Association. They are as follows:

- The decision must also be the patient's own decision.
- The patient's request for physician-assisted suicide must be voluntary.
- The doctor may not suggest suicide/euthanasia as an option.
- The patient must have a clear and correct understanding of the medical situation and prognosis.
- The patient must be facing interminable and unbearable suffering but does not have to be terminally ill.

- The doctor and patient must conclude that there is no alternative acceptable to the patient.
- A second doctor, independent from the first, must be consulted and must examine the patient to confirm fulfillment of the conditions.
- The doctor must end a patient's life in a medically appropriate manner.

In the Dutch legislation, doctors will not be accountable to prosecutors, as they have been in the past. Rather, they will be responsible to a panel of peers including legal, medical and ethical experts. The legislation does not talk of a right to suicide and euthanasia. A doctor, on the other hand, has a right to refuse to cooperate.

The cause of death under the new legislation will be listed as suicide or euthanasia. In the past many suicide and killing interventions were misrepresented as deaths from certain diseases. Most suicide and euthanasia cases have involved patients in a terminal stage of cancer. Many advocates for this legislation, however, think the law is too restrictive. They want suicide and euthanasia available without physician involvement.

Although there is a requirement for patient informed consent, the standards for competency or capacity to consent are amazingly low. Children from 12 to 16 who choose to die have to have parental consent. After 16, they can do so without parental approval. Many persistent questions raised by the legislation center on free and informed consent and the related issue of competency. If adolescents can choose death, what about the demented? What about depressed and schizophrenic patients? What about severely immature adolescents? The idea of providing mentally ill patients an opportunity to commit suicide is shocking to most people and has raised issues which need to be addressed.

Holland has been for centuries the most tolerant and the most liberal nation in continental Europe. What happens in Holland affects legislative initiatives in other countries. In May 2002, the Belgium parliament approved legislation which allows doctors to kill patients, under conditions similar to those adopted by the Dutch legislation. Under the Belgium legislation, a doctor may take the life of a patient under the following conditions: (1) The patient consents to the intervention; (2) the patient is in constant and unbearable pain, physical or psychological; and (3) the patient is terminally ill.

It was no surprise when the French Minister of Health, Bernard Kouchner, called for public debate on the issue. A survey in France had shown a surprisingly wide support for physician-assisted euthanasia. The French survey showed that 38% favor physician-administered active euthanasia in cases of incurable disease and uncontrolled suffering. Legislation similar to that just passed in Holland can be expected soon in all European nations. The French Minister of Health plans to visit the Netherlands to study the legislation.

THE CONTAGIOUSNESS OF SUICIDE

Arguments for and against the legalization of suicide are many. I want to focus on some medical rather than legal aspects of the act. I want to begin by calling attention to the fact that suicide, like many diseases, is contagious. The extent of the contagiousness is astounding and is found in every culture.

A good example is the case of two young Japanese women who committed suicide by jumping into a little known active volcano in the 1930's. Soon after reports circulated, six others did the same, and 25 persons had to be physically restrained. By the end of the year 140 persons had committed suicide the same way. The next year 160 persons did likewise, and 1200 had to be restrained. Barbed-wire fences had to be placed to block entrance to the area. Then entrance to the whole mountain had to be closed. By that time, over 1000 people had imitated the first suicides.

A book by Goethe, *The Sorrows of Young Werther*, became a best seller in the late 18th century. It was about a young man who committed suicide because of a romantic failure. It led to an epidemic of suicides by young men who even dressed up like Goethe's character. The book had to be banned. Any suggestion that suicide is an acceptable way of handling problems triggers its contagious dimension. Extensive publicity, excessive reporting, sensational coverage, romanticizing or glorifying the act, providing descriptions of methods, all these factors intensify suicide's inevitable contagiousness.

In the United States any publicized suicide is immediately followed by copycat suicides or suicide attempts. Great numbers of persons have committed suicide by jumping off the Golden Gate Bridge. The first suicide occurred a few months after the bridge was opened. Following that example, more than 100 persons did likewise. Because suicidal ideation is extensive in today's population, the contagion issue is a serious one.

The fact that most societies have modified some of the harshest forms of prohibition against suicide does not mean that they have withdrawn efforts to discourage it. Around the world, different societies try to deal in a different way with the potential of contagion. After the toughest sanctions against suicide were mitigated, policies still had to be developed in order to control the potential of individual acts to produce social consequences. Even in pagan cultures, sometimes harsh measures were taken to avoid the contagion effect. If women were committing suicide, one policy was to expose their naked bodies publicly. If too many military men were committing suicide, their lack of courage was ridiculed.

The point is that suicide is never the strictly individual act that it is claimed to be. There is no way to justify a strictly individualistic perspective. Suicide affects other people and has important social repercussions. Community law and public morality both have a justified interest in the so-called "strictly private act." Paradoxically, even the extreme autonomists, who argue that suicide is a rational individual option, also make the claim that suicide solves socio-economic problems, especially problems arising when an elderly population becomes burdened with life and burdensome to others.

A PUBLIC HEALTH PERSPECTIVE

Legalized suicide is proposed as a solution to social problems when in fact it is itself a major public health problem. Most cultures today are experiencing an increase in suicide. In the most secular cultures (e.g., Russia and Scandinavia) suicide rates have reached epidemic proportions. In the United States someone commits suicide every 17 minutes. The highest incidences of suicide are among the elderly. The next highest incidences are among persons in midlife. The highest increases of suicide are among adolescents and college age students.

Those most likely to be affected by the consequences of legalizing suicide are the mentally ill: first among them, of course, the depressed. Alcoholics and drug addicts, schizophrenics, patients with personality disorders, and anti-social personalities are almost as vulnerable. Young men isolated in jail are very vulnerable. So are the unemployed, homosexuals and Native Americans.

Among young people, suicide ordinarily is related to romantic failure or social rejection. When young people kill themselves, they often are expressing violence against someone else. Suicide is a way of "paying others back" for rejection or mistreatment. However, once suicide is legalized for the elderly or infirm, more and more young people will copycat the act in order to "solve" their emotional upsets.

Suicide among young people in the United States tripled in the last half of the 20th century. It is the third leading cause of death among young people in general; it is the second leading cause of death in college students. One in ten college students seriously considered suicide during the last few years; most of that 10% had actually drawn up a plan.[149]

The high school population is even more worrisome. One in five (20%) has seriously considered suicide. Most of the 20% have gone so far as to draw up a plan. One in ten actually attempts suicide. These figures were gathered in 1997 and are substantially the same as those gathered in 1993 and 1995 surveys.[150] The figures are higher in places like New York and Oregon where the Federal Courts where assisted suicide is legal. European and Canadian statistics are somewhat lower but not much. The same is true of statistics from Latin America and the Caribbean. More teenagers die from suicide than from cancer, heart disease, AIDS, pneumonia, influenza, birth defects and stroke combined.[151]

During the 1960s and early 1970s, the U.S. society was in turmoil over the loss of young lives in the Vietnam War. That war created a social revolution and a hippie culture which are still with us. And yet, when the data is examined, there were almost twice as many deaths from suicide among young people (under 35) during that period as there were deaths from the war. Figures from the 1980s and 1990s show that more young people died from suicide than from AIDS. People knew about the Vietnam deaths when the war was raging. And they learned about death from AIDS and HIV through the media. But the number of deaths from suicide is not widely known, and therefore the social consequences are likely to be ignored.

In other words, the public health aspect of suicide is commonly overlooked. People do not think of the fact that suicide, exercised as an individual liberty and equality right, would spread the practice first among the elderly and then among young people. Legalized suicide would publicize the justifications, and consequently, increase the incidence. Legislators need to consider information and statistics from medicine and from the social sciences when they propose to write new laws.

One U.S. senator, Harry Reid, whose state of Nevada has the highest suicide rate in the nation, recognized the problem. He offered no change in legislation but at least got passed a resolution that named suicide as a social problem. His resolution also outlined a set of strategies for suicide prevention and for the treatment of mental disorders which cause it. It will be interesting to see how the debate plays out between the public health perspective and the one which sees suicide as an individual right. If the autonomists win, responsible leaders will have to pay careful attention to the effect of their new legislation, especially on the younger population.

Is the criticism of physician-assisted suicide which focuses on contagiousness and social consequences an alarmist reaction? Those who oppose legalization argue that patients asking for death should trigger concern about depression on the part of health-care professionals. They want better care of the dying. Those who support the law insist that legalization of physician-assisted suicide will create an environment in which people can more easily talk about dying. Is it unlikely that serious ill effects will follow from legalizing suicide? We have some data from Oregon, Washington, and Montana. The reasons given by most of the persons in the states who chose suicide were (1) they felt that they had become a burden to others; and (2) they wanted to have control over their situations. So far, the legal suicides in the states have been few, and there is no data on copycat suicides. Proving contagion will be difficult because of the lack of data or inherent ambiguities in the data. As of December 2002, fewer than 100 people in Oregon had died from assisted suicide.

Most data comes from Holland. The practice has widespread acceptance in Dutch culture. Legislation legalizes both physician-assisted suicides and active euthanasia. Will young people there be in greater danger as a result of formal legalization and widespread cultural acceptance? Statistical data over the years will be important.

Recently a how-to guide of suicide methods appeared on a Dutch website. Now depressed people have easy access to information about how to carry out the act, and so do young people. How much will both groups be influenced by the suicide instructions? The background music on the website is "Goodbye Cruel World" by Pink Floyd. The site is called, "Thisbe's Self Destruction Site." (Thisbe is a mythological character who committed suicide.) The author of the site said, "I refuse to accept any responsibility for the consequences of putting to use the things I have written." Suicides by young and depressed people are already increasing; and given what we know about contagion, the overall effects of legalized suicide on the whole of Dutch society will not be able to be ignored. Recent street demonstrations against the Dutch law may signal the fact that some people see the need to take responsibility for what is happening.

A PSYCHIATRIC PERSPECTIVE

Suicide is the common end point of many different forms of severe psychiatric illness. The combination of mental illness and upsetting circumstance does not necessarily cause suicide, but the two factors are almost always present when suicide occurs.

The most common mental illnesses related to suicide are mood disorders: depression and bi-polar disease. Other mental illnesses linked with suicide are schizophrenia, borderline personality disorders, antisocial personality, alcoholism and drug abuse. The link between these mental illnesses and suicide is so strong that rarely does one find a suicide without clear indications of mental pathology. Even the most painful, disfiguring, and dangerous physical disorders are not accompanied by increased suicide rates (e.g., MS, Huntington's, cancer) Some physical diseases do cause bouts of depression, but it is the depression that is most obviously operating when sufferers of those diseases take their own lives.

What is true of physical diseases generally is also true of the physical diseases of elderly people. Even when the elderly are infirm, and suffer from one or more debilitating illnesses, it is not the physical disability that leads to suicide or suicide attempts, but the depression which accompanies these illnesses. The idea of making help with suicide a basic right for the elderly while leaving mental and emotional illness in the elderly unattended ignores the medical facts and guarantees social disaster.

If suicide is the product of severe mental and emotional illness, then not surprisingly, suicide notes which people leave most often reflect the deep despair and the total loss of hope. Suicide notes show the signs of severe depression. Some patients are intelligent and have insight into their disease. Some also have the ability to verbalize their dreadful situation. Sometimes, not even love and good family relations can counter the pain of severe mental and emotional illness. The idea of assuming that suicide is most likely to be a rational act is unreasonable. The person requesting suicide help is more likely to fail the competency test.[152] Suicide is the final expression of serious illness. Difficulties alone don't cause suicide but serious mental illness does.

No doubt life problems are associated with suicide. These may be cited as the justifying reasons for the act: alcoholism, drug addiction, disappointed love, betrayal, divorce, family discord, financial distress, job loss, criminal indictment. These and similar difficulties can trigger or aggravate psychiatric illness. The relationship between life's difficulties and psychiatric illness is a very complex one. We do not always know which comes first or which one aggravates the other. Common sense, however, suggests that psychiatric disease, even before its full-blown expression, is like an impaired immune system. A life difficulty which one person can pass off collapses a more vulnerable person prone to mental illness.

Severe depression is the most serious of all psychiatric illnesses because most commonly it leads to death. Depressed persons focus on negative aspects of life. Any sort of event is likely to trigger off deeper depression. The deeper the depression the worse the hopelessness and the more likely it is that the patient will attempt suicide.

Thirty thousand Americans kill themselves every year, and nearly one half million attempt suicide and need emergency treatment.[153] If the Supreme Court had declared suicide a constitutional right, we can only imagine the result on society, especially on young people. The numbers of people taking advantage of this "right" probably would have skyrocketed. Given the well-established history of one suicide leading to others through imitation, we can easily imagine moving swiftly toward a suicidal culture.

Rather than legalizing suicide, isn't it more reasonable to ask for better treatment of mental illness for elderly patients? Much of the pain and suffering which precedes suicide can be relieved by physicians. The same may be true of the pain and suffering of family members following a suicide. What seems undeniable and inescapable is the fact that suicides lead to more suicides.

The present campaign for legalization will change the way life as well as death is experienced in the United States. Aging always involves anticipation of death, but death need not be caused by mental illness. Aging will be a radically altered experience if it is turned into an anticipation of suicide. Aging can be made a more humane experience in societies where available medications for emotional illnesses are part of long-term care programs.

CONCLUSION

A recent study reported on the evening news showed that suicides are on the rise. Already they surpass the human lives destroyed by murder. One might expect a liberal Catholic to support a person's freedom to choose suicide, but some personal choices, even if they are free, have heavy negative social consequences; and human beings are social animals with social responsibilities. In Catholic ethics, freedom is important, but it is not the only value. Life counts and so does community responsibility.

The argument being made here is an example of a liberal perspective supporting a traditional religious teaching. The tradition enshrines an overall vision that distinguishes between individual morality and a social ethics. A liberal Catholic perspective insists upon humane medical practice and the influence of basic Natural Law principles on private life and public law. A Natural Law perspective supports laws which proscribe the direct and intentional taking of human life. It supports the distinctions which are part of the traditional biblical moral teachings. Paradoxically, it is conservative Catholic moralists and Catholic bishops who sometimes reject the traditional distinctions between ordinary and extraordinary treatments and between active and passive euthanasia. The Archbishop of St. Louis, for example, tried to force futile and burdensome interventions on a dying patient. His rejection of a humane and reasonable Catholic moral tradition caused widespread shock and scandal.

But what if pain relief is not effective and nothing can be done medically and a person is dying in agony? These rare circumstances might create a justifiable exception to the rule against direct and intentional killing. I already referred to the medieval practice of directly killing a soldier who is dying in agony. If a patient is dying and in agony and there is neither medical hope nor adequate pain relief, a case can be made for a private act of mercy in such a situation. This circumstance or situation would not make the act of killing right or good. This rare case would not be the basis for changing legislation or creating a legal right to be killed. But it does create an extreme case, and I believe a person using his or her own individual moral judgment, might privately justify a mercy killing.

That's one thing. That is an example of tragedy or a tragic dilemma. In a moral dilemma whatever one does would be wrong. I wish I could argue a case which covered every possible condition in life, but I can't. So I end by admitting that the Judeo Christian tradition, the Hippocratic tradition, the Western medical tradition are all reasonable and respectable but may have a limit. If so, I've suggested a way of handling it: i.e., mercy killing, for which one is personally responsible and vulnerable to prosecution. Not enjoying legal protection forces careful consideration on the part of the physician assisting suicide or euthanasia. This is a long way from the campaign to make killing a new right. Legalized killing threatens the lives of the most handicapped, the poorest, the most deprived, and the least cared for.

XII. Palliative Care For Dying Patients

INTRODUCTION - THE THEORY BEHIND MODERN MEDICINE

Medical practice is always the expression of a theory of medicine. Before contemporary medicine with its particular background suppositions, a different theory and practice operated in the West for more than two thousand years.

Greco-Roman medicine used a Hippocratic or Galenic theory, based upon the relationship of elements of nature (fire, air, water, earth) or humors of the human body (blood, phlegm, yellow bile, and black bile). Health was understood in this theory as harmony or balance of the humors. Disease was understood as imbalance or excess of one or another component. Medical practice was directed toward reestablishing balance either through the *dieta*, a pharmacon, or surgery. Traditional medicines generally were holistic, and ancient Greco-Roman medicine was no exception.

The background suppositions which replaced the humeral theory began to emerge in the Renaissance with the beginnings of modern anatomy, physiology, pathophysiology, and finally microbiology.[154] By the late nineteenth century modern science and rigorous laboratory methods became the operative background theory of mainline western medicine. Medicine in the 20th century became scientific medicine.[155] Patients were looked at through the lens provided by science, i.e. as physical body parts rather than as whole persons. Holistic medicine today tends to be alternative medicine.

Modern medicine is German medicine in its origin. It is grounded on hard scientific data and rigorous scientific procedures. It focuses on physical body parts (organs, tissues, cells) that are diseased. Practitioners using this vision or theory become experts on some particular diseased part (e.g. nephrologists, urologists, cardiologists, dermatologists, orthopedists, gynecologists, podiatrists, pulmonologists). Specialists identify disease and then use powerful drugs and technologies to eliminate it. The person of the patient is reduced to his or her physical pathology and then aggressively treated. The goal of modern medicine is to eliminate pathology in specific parts and this goal is pursued until the very end of life. Integrated with this narrowly focused contemporary medical theory is a military metaphor, which explains the way modern medicine is practiced. *War* in this military metaphor means all-out war rather than a just war theory. (President Nixon declared a war on cancer.) A patient's death means that the war is lost. Disease is understood as the *enemy* and physicians do *battle* against disease. Doctors order *batteries* of tests and look for magic *bullets*. They think of themselves as being on the *firing line* or in the *trenches*. Cells can become *invasive* and *attack* the body's *defense system*. The top doctor is a *surgeon general*.

To this narrowly focused militaristic style of medicine, is credited the many new effective interventions which are sought after by sick persons everywhere in the world. In the U.S., television is inundated with pharmaceutical ads in which some patient makes the claim: "modern medicine saved my life." Certainly modern scientific medicine has saved lives and extended many others. But there is a downside to this powerful modern medical theory and its corresponding style of practice.

When the aggressive and narrowly focused modern medicine continues to be applied at the end of life, it can cause something horrific. In fact, sometimes the consequence can be so horrible that those who see it come away saying, "I don't want to die like that" or "I never want to go through anything like that." The concentration on curing diseased parts hides the needs of the whole person – especially the need for care, accompaniment, pain and suffering relief at the end. The focus on curing pushes aside caring which doctors and nurses could provide if only it were seen as a valid medical objective. The socio-political movements in Europe and the U.S. to legalize suicide and euthanasia are one response to the failures of mainline medicine to address the inevitability of death and dying.

In Europe legalization of physician-assisted suicide and physician administered euthanasia is spreading. Formal legalization occurred in Holland in 2002, but both types of physician assisted taking of life were decriminalized in the 1980's and have been widely practiced for decades. Belgium is preparing to legalize the practices and pressure is being felt in other countries to do likewise. In the United States, individual patients demand the "right to suicide" and many legislators advocate for legalization of the right. Oregon may be the only state where physician-assisted suicide is legal, but essentially the same legislation is under consideration in other states. If successful, the movement to legalize physician-assisted suicide and euthanasia will reverse a long ethical tradition in the West. By every indication, the movement will continue to gain momentum unless pain and suffering at the end of life come to be adequately addressed in mainline medicine.

PALLIATIVE MEDICINE: AN ALTERNATIVE VISION

Dissatisfaction with the practices of mainline modern medicine, especially at the end of life, explains the expansion of hospice, as well as the development of a new medical specialty called palliative care. In both these new movements, the focus is switched away from aggressive attack on body parts and defeat of the enemy death. Their focus is on the whole person and on the quality of a dying patient's life. Their background medical theory or vision is more holistic. Their objective is to create a more humane medicine.

In Hospice, death is recognized as inevitable and the focus is on making care for the dying more humane. Hospice is about caring rather than curing. A whole team of hospice caregivers gather around a dying patient to address not just problems with distinct body parts, but broader problems of pain and suffering. Hospice caregivers are concerned with social, psychological and spiritual needs of the patient. The environment created around a dying patient changes from being a technology-dominated battleground to being a warm, caring, loving place for both patient and family.

Palliative Medicine adopts a background vision similar to Hospice. When the focus of a medical practitioner is on specific body parts and on the conquering of disease, pain and suffering are given far less importance. Pain for the mainline physician may be understood as pointing to disease and therefore helpful for diagnosis and treatment. If so, removing it is seen as a mistake. Since pain and suffering cannot be measured in a completely objective way, they are inevitably unclear and somewhat ambiguous. Consequently, they fall outside the narrowly scientific lens. In addition, since pain-relieving drugs are addictive, dispensing them can be considered dangerous.

Palliative medicine moves away from this vision and the practices which it creates. Aggressive treatment of scientifically accessed pathologies is no longer the only goal of medicine. For the palliative care physician, another goal is the relief of pain and suffering. Palliative care does not reject modern scientific medicine but broadens the background medical theory in order to correct some of its worst limitations.

To palliate means to mitigate, alleviate, lessen, abate. Palliative medicine pursues three separate but related objectives:

1. Effective focus on and management of pain and suffering
2. Concern for both the bodily condition and for the inner life of the patient
3. Decision-making which respects patient autonomy and the role of legal surrogates.

We will be addressing these objectives by focusing on pain, suffering, and patient decision-making at the end of life.

PAIN

Pain is a negative feeling which originates in some damage to the body. It is the neural transmission of a noxious stimulus and inevitably it is a part of bodily existence. Usually pain accompanies our entering into and our exit from life. In order to provide effective pain relief, medical professionals have to understand pain in all its complexities and the human nociceptive apparatus is tremendously complex.[156]

The nervous system pathways transfer signals from an injured body part to the brain. The nociceptive nerve fibers make connections with spinal nerves that ascend to the thalamus and from the thalamus to the cortex. The neurological signals can be diminished or enhanced, decreased or increased by what happens in the cortex. In other words, noxious sensations can be influenced by the patient's understanding of what is happening. Neural pathways from cognitive centers can modulate or even inhibit pain transmission from below. Besides the neurological pathways there are cellular and chemical components to pain. Endorphins, for example, exert analgesic effect on the nervous system.

In order to provide effective pain relief, medical professionals have to be able to manipulate the neural, chemical and cognitive dimensions of a pain experience. They have to be able to address issues of anatomy, physiology, and psychology. They have to address patients as persons. Unconscious patients or patients in a vegetative state may show neurological reactions to noxious stimuli but do not have what we call pain because they lack the cognitive component of pain. Pain is a term which refers to a complex process which involves both sensing and interpreting, neurological signaling and assigning of cause, corporeal and mental activity.[157]

RELIEF OF PAIN

Pain exists when a person says, "It hurts." There is a physical or somatic dimension of pain, but pain is not reducible to a neuronal or somatic reality. Pain is an experience which joins together different symptoms rather than a single phenomenon. It is a particular individual person's experience. It is distinct and varied.

Although pain may raise blood pressure, the pain experience cannot be grasped in purely quantitive terms. The experience cannot be fully accessed by scientific instruments. Pain is not measured the way blood pressure, or blood sugars, or heart rate can be measured. There are widespread bodily responses to pain, including changes in blood pressure, heart rate, breathing and sweating. But these measurable changes are not the experience of pain. Pain subtleties cannot be reflected through objective measurements. Ordinarily a physician has to depend upon communication with the patient about the pain and has to trust a patient's report.

Some palliative-care physicians focus primarily on blocking the neurological transmission of noxious stimuli with analgesics. Others focus more on the involvement of the brain and the central nervous system. Many patients with pain problems who have no access to palliative care physicians wind up transferring to alternative medicines and alternative therapists who ordinarily are more responsive to patient experience and more holistic in their strategies of intervention.

The experience of pain puts both doctor and patient in contact with the complexity of human life. In pain, we humans face the age-old problem of how something physical becomes a feeling, a thought, an experience, and then how in reverse, thoughts and feelings and experiences can have an influence on neurology and physiology. Because pain is a human experience it can never be reduced to pure physiology. The physiology however is so complex that effective pain mitigation requires an extensive grasp of the nociceptive apparatus and of the varieties of medical interventions to alter its signals.

To say that pain cannot be understood the same way blood pressure can does not mean that it is beyond all quantification or measured understanding. Different modifiers, which are used to describe pain, reflect some of this understanding. "Chronic" pain lasts more than six months and is distinguished from "acute" pain. Neuroendocrine components (endorphins) can modify the nociceptive apparatus and help to explain pain which endures beyond the initial tissue damage. The tissue damage itself may endure (e.g., rheumatoid arthritis), but pain may also endure psychologically (e.g., when pain remains in a severed limb). Emotional factors (worry, anxiety, stress) may increase pain and cause continuing pain.

The psychological and emotional dimensions of pain are evident in the attention which is ordinarily paid to pain or in a distraction which can liberate from pain. In an emergency, a person's attention may be fixed on escaping from the danger. The pain from a wound which was inflicted appears only when attention is refocused on it. Pain has both physiological and nonphysiological dimensions.[158]

The most important non-physiological component of pain is its meaning. Pains are experiences which mean something to human beings. Pain in the chest for me, with my medical history, means something different than it does for the Brazilian goalie who uses his chest to block a shot in the world cup game. Pains are evaluated experiences. Pain which a person believes comes from cancer is experienced differently from the same physical experience evaluated as heartburn. One signifies death, the other a need for Tums.

Every reality for human beings suggests a meaning, and the assigned or perceived meaning influences first how pain is experienced and then how it will be responded to. There are internal as well as external, subjective as well as objective, mental as well as physical dimensions to the pain experience. The word used most often to refer to the internal, mental, subjective aspects of pain is suffering.[159]

SUFFERING

Pain and suffering ordinarily go together, but the different words sometimes refer to different realities. Depending on how it is understood, the same physical pain can cause either overwhelming suffering or almost no suffering at all. The pain in the chest caused by blocking a game saving shot in a World Cup match may even be a delight, while the same physical experience in a spectator with a history of heart trouble may cause suffering to the point of panic. The experience of suffering sometimes can be radically altered by a change in the meaning or the purpose of pain and suffering.

This is an important consideration for professionals who provide palliative care for dying patients. If the palliative-care physician can effectively address a dying patient's physical pain and do so in such a way as to assure the patient that the pain can and will continue to be controlled, then the meaning of the pain changes and it becomes much less of a suffering experience. Once a dying person is assured that his/her physical pain can and will be controlled, the suffering as well as the pain is diminished. Suffering is very much involved with worry and worry is a mental and an emotional thing which implies a sense of the future.

If patients lack any sense of future control over their condition, the suffering associated with their pain will be much greater. Even after a particular physical pain has been relieved, dying patients may continue to suffer from the worry that the pain will return and will not be relieved. Typically, migraine headache patients suffer from fear of migraines even when they are without the actual pain.

The body produces the anatomical and physiological grounding for a pain experience; but thoughts, memories, hopes, feelings, beliefs and meanings are the main sources on suffering. Suffering most often comes from loss of hope, loss of purpose, loss of meaning, depression, despair, existential distress. For suffering in the dying process to be controlled, dying patients need compassionate attention to their mental state. They need assurance that they will not be abandoned and that their worries and anxieties will not be ignored. In order to provide palliative care, physical pain must be relieved; but suffering too must be relieved, and this requires a broad spectrum of human care.[160]

Where a loving family is missing and a patient is dying, then palliative care providers must take their place. Full palliative care requires a physician skilled in physical pain relief plus a team which supplies the physical, psychological, social and spiritual support needed to manage the suffering which all too often accompanies descent into the dying process. Bodies produce the pain experience. Only persons with minds and souls, with purposes and meanings, convictions and future expectations suffer.

RELIEF OF SUFFERING

Suffering occurs when dimensions of one's internal self and personal integrity are threatened. Relief of suffering can only come from addressing these threats. A dying patient may not be in pain biologically but yet be suffering terribly because of a threat to his/her inner self and integrity. Persons cannot properly be cared for in palliative medicine by attending only to corporeal considerations. Persons who are suffering are saying that their whole life hurts. Suffering is an existential experience and relief starts with compassionate attention to the whole person.[161]

The human being is an integrated whole, composed of different dimensions. The internal or subjective dimensions are pulled together under concepts like self, soul, mind, character. Suffering can originate from the physical parts or from the spiritual self. Suffering always accompanies loss or threatened loss of any dimension of the whole person.

The subjective components of a particular individual taken together are referred to as personal integrity. Among these dimensions are the following: conscious mind, the unconscious, beliefs held with commitment, family experience, culture and society, associations and relationships, a sense of future, meanings, purpose, attitudes and dispositions, sense of stability and continuity, willpower, control of self and environment, a personal story.

Suffering is rooted in a loss of any one of the dimensions of personhood. Providers of palliative care must be sensitive to the complexity of humanness and be willing to try to relieve suffering related to any dimension of the patient. Aging and dying are all about loss of those realities which identify and define us as human beings. Loss threatens a person's self and integrity.

Because suffering is related to realities of each unique person's self, it is always individualized. The same physical pain can cause different suffering in different persons. The source of a person's suffering is accessed through personal communication rather than scientific tests. The suffering person may be primarily concerned about a physical pain, but the provider of palliative care will have to be sensitive as well to different threats to a patient's inner life and personal integrity. It is a challenge to gain an understanding of what it is that is causing a particular person's suffering. The sufferer himself or herself may not know.

Maybe the patient has a sense of failure. Maybe a patient has failed in his or her life expectation. Maybe there is a sense of having failed the expectations of others. Besides feelings of failure, there are feelings of anxiety, anger, loneliness, guilt, unfairness, abandonment, which may be causing the suffering.[162] The same feelings may cause suffering in one person but not in another. Sometimes pain will cause suffering, sometimes not.

The human person is not reducible to a body, but bodiliness is part of personhood and bodily disintegration and physical loss may be a source of suffering. Loss of bowel of bladder control is a common cause of great suffering. People usually want to live and they fight the dying process. This struggle is a possible source of suffering. The struggle to keep one's personal integrity in the face of bodily disintegration is a major challenge for the dying. Helping dying persons to do so is one more major challenge for providers of palliative care. Palliative caregivers have to be concerned with a patient's bodily disintegration. Said differently, to relieve suffering they have to attend to the physical needs of the dying person.

Sigmund Freud and other great psychologists spent lifetimes trying to understand the non-physical dimension of human life – the inner self. If no inner self can be fully understood, and if fears of loss of one's inner self is a source of so much suffering, then providing palliative care for the dying is understandably challenging. What aspects of a complex inner self are more threatened in a particular dying patient: the person's purpose, his meaning system, her beliefs, family concerns, identity, integrity, sense of continuity, security, self control? How can the threats to one or all of these, and the corresponding suffering, be relieved? The basic questions facing the palliative care provider are two. Is this person in physical pain? Why is this particular person suffering?

TWO IMPORTANT SOURCES OF SUFFERING

1) Sense of Purpose

One critical component of an inner life is a person's sense of purpose. One person's purpose may be to preserve himself/herself for as long as possible. Another person's purpose may be continuing evolution and self-development. The purpose of some modern secular persons may be continuing accumulation of wealth. A different purpose would be to testify to one's religious faith or to leave a legacy of faith for one's family. A sense of purpose is part of a person's self.

The physical parts of life incorporate purpose. Legs move, hands grasp, eyes see, brains make it possible to wonder and inquire and understand. Loss of physical function can be the source of both pain and suffering. A whole person has purpose, and that purpose will help define each person's self. Pain can interfere with an individual's purpose, but loss of purpose can cause the very inner self of a patient to disintegrate. If this happens, terrible suffering can result. Even decline in activity may contribute to a loss of purpose. Since dying interferes with all human activity, its anticipation often threatens one's sense of purpose. Consequently, it will cause suffering which must be taken into consideration by palliative-care providers.

Think of the inner purpose of a great scholar whom you may know and the associated academic activities. Disease may interfere with his/her self-designed purpose and consequently cause terrible suffering even without corresponding pain. Suffering may be overcome temporarily, even with pain and illness, when purposeful activities are able to be resumed.

If dying comes slowly, the major suffering may come from the inability to carry out one's self-defining purpose. Even the realization that one's purposeful activities soon will be lost (a sense of what the future holds) can cause suffering. The dying process with the accompanying disabilities and dysfunctions can focus attention on the distressed body and away from life's purposes. But the opposite is also possible. Death and dying can be understood as a part of life and not inimical to life's purpose. Palliative caregivers have to be attentive to a dying person's life purpose and be sensitive to the possibility of loss of purpose causing suffering at the end.

2) Place of Meaning[163]

Associated with purpose is the place of meaning in human life. Meaning we saw plays a role in the experience of pain. Meaning, like purpose, is essential to personhood and if it is lost or threatened, so too is personal integrity. Pain and disease and the dying process may threaten beliefs that earlier gave meaning to a life. Sometimes the worst suffering at the time of death can come from the threat to a person's meaning system. "Why is this happening to me? Why is there such evil in the world?" Job suffered from just such questioning. Why, he asked, were awful things happening to him? How could he reconcile such things with his belief in a just God? The questioning caused suffering which was finally relieved by a reformulation of and a recommitment to his religious meaning system. Persons who provide palliative care may have to address this philosophical dimension of personhood in order to relieve patient's suffering. Palliative caregivers may have to be philosophers of a sort. Sometimes they may have to be priests, because a person's deeply held core inner meanings may come from religion.

If suffering at the end is related to a loss of meaning, then palliative care may come from help in a recovery of meaning. For dying Christians, linking a patient's suffering to Jesus' suffering at the time of his death may help to restore meaning and hope. Even Jesus however suffered at the end. Even His inner integrity was threatened. Listen to his cry at the end: "My God, My God, why have you abandoned me?" Providing palliative care for a patient may involve helping the dying patient to retain a sense of personal integrity and inner wholeness by connecting the dying process to a meaning system.

MEETING THE CHALLENGE OF PALLIATIVE MEDICINE

In order to relieve pain and suffering the palliative-care professional must first understand the complex reality of human personhood. Besides being an expert in neurology and the subtleties of pain transmission, the palliative-care physician must understand the complexities of acute and chronic disease. Such understanding is critical for deciding whether to intervene and hoe to intervene with patients at the end of life. Relief of pain and suffering (rather than systematic diagnosis and aggressive treatment of disease) is a new goal of humane medicine.

In mainline contemporary medicine, science is used to define disease and to develop effective therapeutic interventions. Science is the lens through which contemporary physicians understand and treat patients and with this lens the inner self of a patient is out of view and therefore unattended. With some ambiguity, pain may appear on the scientific screen, but not suffering. Consequently, contemporary mainline medicine all too often is impersonal.

The new specialty of palliative care breaks with this defective perspective, and re-focuses on the whole person of the patient. Pain which was marginalized and suffering which was ignored again become central concerns. Palliative-care medicine and the hospice model of care correct a serious deficiency and give modern medicine a needed broader vision.

Looking back in time we find many supports and confirmations for this revised medical vision and it's objective of providing a more humane dying. The Talmud,[164] for example, the authoritative source of Jewish moral teaching, provides many hospice-like directives for caring for dying patients. Another unexpected source of confirmation for this new medical vision is Leon Tolstoy in his novel, *The Death of Ivan Illych.* Both Tolstoy and the Talmud emphasize the importance of staying with a dying patient in a compassionate way. Both emphasize what we have spoken of as relief of pain and suffering.

The suffering of Ivan Illych as he lay dying was horrific: a testimony to how ignoring a dying person's inner self can intensify suffering to the point of torture. Who provided palliative care to the subjective inner self of Ivan Illych according to Tolstoy? His servant did. How did he do so? By staying with his master, physically and psychologically, by showing compassion, by addressing the main source of his suffering which was his terrible loneliness.[165] In contrast, Ivan's family abandoned him physically and emotionally. They stood by but engaged only in inane babble.

There is no cure for dying; but the best, the safest, the most certainly effective palliative care is provided by physical pain relief and by staying with dying patients in a perceptive and compassionate way: holding their hands, rubbing their foreheads, massaging their legs, by listening and speaking to them softly, and all the while being aware of the patient's inner self and personal story. Protection against suffering from despair at the end of life comes from being able to look beyond death. To be able to look beyond death with faith and hope toward a future is most likely in the presence of caring and compassionate others. Being with the dying person in an understanding and compassionate way is the best way to relieve of suffering and the safest kind of palliative care.

PALLIATIVE CARE AND BIOETHICS

In the ethics of contemporary medicine, patient competency to give free and informed consent to every intervention is crucial. Patients participate in contemporary medicine by becoming informed about their condition and about the potential risk/benefit ratio of proposed interventions. Then the patient freely chooses to be treated or to forego treatment.

At the end of life it is particularly important to provide truthful information to patients. Patient participation in medical decision-making requires that the patient first is able to understand his/her medical situation and able to understand the advantage or disadvantage of proposed interventions. Then the patient or the patient's surrogate has to be able freely to choose for or against an intervention. The contemporary medical ethics of informed consent presupposes a patient with capacity to understand and freely to choose. Typically, this ethic does not take into consideration the impact of pain and suffering in the dying process.

In order to give free and an informed consent, patients must be self-directed. They have to be able to reason about their situation. Especially at the end of life patients have to be reflectively aware of who they are, their life goals, their family interests, their meaning system, their sense of purpose, etc. To reason about and then freely decide about an important medical issue requires that a patient take his/her complex inner self and life experience into account. But pain and suffering may upset this type of reasoning and decision making. Suffering especially tends to direct reasoning and decision making away from prudent evaluation. It creates pressure, coercion, and compulsion to move in some one direction and little else. Consequently it can undermine a patient's competency to make free and informed medical decisions. Surrogates frequently have to make decisions for dying patients.

If pain and suffering have a negative effect on the competence of dying patients to make free and informed decisions, then palliative care can help to restore a patient's competence. Hospice and palliative-care professionals may have to evaluate a patient's competence to make medical decisions in dying situations. The palliative-care professional may have to help the suffering patient or the patient's surrogate to decide. The palliative-care professional may have to work to break through the isolation created by pain and suffering and to make sure that the patient does not make decisions which are harmful or contrary to his or her loss. Pain and suffering can obscure patient benefit/risk. Therefore, the palliative-care physician, especially at the end of life, may also have to be a bioethics consultant. In cases where there is no family or surrogate decision maker, the palliative-care physician may have to choose what is most humane for the incompetent dying patient.

TECHNOLOGY

XIII. An Ethics of Technology for the 21st Century

INTRODUCTION

The judiciary is the quietest of the three branches of U.S. government. This is especially true of the highest level of the judiciary, the Supreme Court. Except for opinions written most often in dense "legalese," we hear very little talk from Supreme Court judges. Most Americans know little or nothing of the judges as persons.[166]

A change in this pattern is occurring with one member, Supreme Court Justice Stephen Beyer. He is going around the country speaking about his personal concerns and the dangers we all face coming from genetic science and technology. He makes the point in his talks that the judiciary will be called upon to respond to aspects of the revolutionary developments in these fields. Consequently, he asks for dialogue to begin between those involved in genetic science or technology and the broader community. His point is that we will all be affected by the changes which are occurring; therefore, we all need to be involved first in becoming better informed about what is taking place and then in making decisions about what is right and wrong.

Judge Breyer is primarily focused on the fact that judges will be called upon to make decisions about gene patenting, gene testing, stem-cell research, patient privacy, etc. His interest is in encouraging communication between scientists, ordinary citizens and the judiciary about these matters. His point is that judges have to learn about what is going on in contemporary science and technology so that they can make reasonable decisions for all of us about legal issues generated by the science and its concrete technical applications.

But what is true of judges is also true of medical professionals and hospital administrators and persons in Catholic health care responsible for ethics. It is equally true of religious and political leaders who have to provide moral direction for ordinary Church members and ordinary citizens, all of whom have stakes in the consequences of today's science and technology. Every issue which the judge calls attention to is both scientific and ethical. Supreme Court justices certainly have a responsibility to learn about contemporary science and technology in order to form opinions and make legal decisions. But so do persons in responsible health-care positions in Catholic health care. Hospital ethics policies and Catholic moral teachings, have to be responsible, which means they have to be based on in-depth understanding of the realities of science and technology.

What follows tries to make a case for cross-disciplinary and inter-religious communication and for serious attention to developments in contemporary science, technology and ethics. I does not presume to provide final answers, but rather attempt to start an ongoing dialogue between scientists, biotechnologists, bioethicists and health-care professionals. Dialogue in the liberal Catholic tradition has to take into account the reflection and insight that have already taken place on these topics in history.

HUMAN BEINGS AND TECHNOLOGY

Human animals are technological beings. Ordinarily, engaging in this distinctly human activity creates advantages for human development. This was obviously true when humans fashioned hunting technologies which created more food of better nutritional quality, along with more protection for the hunters. Later, military technology made for more efficient killing and consequently greater survival advantage in the never-ending battles with other humans. Then, technologies moved humans from hunting to farming and again created better food and more reliable nutrition. The agricultural technologies contributed to the development of written language, mathematics, literature, science, and of civilization.

Years later, technological developments created the Renaissance in Europe and then the Industrial Revolution. Machines of mass production suddenly overpowered hand-driven arts and crafts, and a modern culture began to emerge centered around shopping and the accumulation of things. The same advances in understanding which made possible so much of modern technology and modern culture then put both in danger of extinction by producing weapons of mass destruction: nuclear and hydrogen bombs, deadly gases and biological weapons.

One thing that has been obvious from the beginning is that both good and evil consequences come from technologies. Every technology thus raises questions about ethics. What is different today is that the ethical questions have become more pressing because the potential for good and evil consequences has expanded. On the good side, for example, we see the possibility of curing terrible diseases, improving human existence, and expanding food supplies sufficiently to relieve widespread malnutrition. At the same time, and from the same technologies, we see the possibility of creating devastating new diseases and actually obliterating the food supply.

Today, we find ourselves at the start of what appears to be yet another technological revolution, one in which powerful technologies will be in the hands of almost everyone. Since so many people will be affected by today's technologies, shouldn't ordinary people have some say about their approval or disapproval? But what standards will be used? Now, as never before, there is reason to begin to think seriously about an ethics of technology which would help to advance the good and to repress the evil.

The most widely appreciated medicines (penicillin) have now created antibiotic-resistant bacteria that threaten uncontrollable deadly infections. Respirators and feeding devices, which were designed to support life until recovery could occur (half-way technologies), now are being used to extend the dying of patients both young and old. Reproductive technologies can now relieve infertility for some couples, but also threaten the natural procreation system. Even at their best, they also frequently result in premature births and multiple pregnancies that a woman cannot bear and a couple could not care for. The eight babies born to a single mother after reproductive technology clearly illustrate this point.[167]

NEW TECHNOLOGIES AND NEW EMPHASIS ON ETHICS

The ethical issues generated by technologies can ordinarily be somewhat satisfactorily addressed by using a utilitarian perspective: i.e. by weighing risks and benefits. Such weighing is always problematic, but at least this approach for making ethical judgments about limits is fairly clear-cut. This is not the case with all technologies. Today much more powerful genetic technologies have emerged which can bring about another eugenic movement with all its devastating effects on human life. The ethical issues generated by genetics and genetic engineering are many: the ownership of genetic discoveries, patenting, privacy, discrimination in insurance and employment, testing, counseling, somatic cell and human germline therapies, eugenics and cloning. The last two are especially worrisome because their risks and consequences can be enormous.

It is one thing to use technology to intervene into genetic structures in order to cure a disease. Risk and benefit weighing is an appropriate ethical approach for such decisions. Intervening in order to enhance certain traits or to create an "ideal" population is different. Now instead of weighing the consequences of the act, the focus of ethical evaluation has to be on the act itself. Would such a use of technology ever be justified? Do such interventions reduce human beings to objects? If so, they violate the foundation of a Natural Law-based ethics: the structure of human life or the created human order.

The impulse to enhance individual lives and even whole populations is not new. Most recently the Nazis tried to do both genetically, and the consequences were horrible. A repugnant and destructive background racism seems to accompany eugenics movements. The same flawed motivations for enhancement appear again and again, followed by the same terrible results. And yet the enhancement impulses are still with us. Some people are anxious to take advantage of the latest genetic technologies to produce whiter or taller, more athletic or more artistic "products." Could such technology be justifiable? Or is the very act of using eugenic technology in order to create particular human products a violation of human dignity?[168]

Eugenics today would involve some form of germline genetic engineering, i.e., altering the genes of sperm or ova so that the intervention would have an impact on future generations and not just on a particular individual. With somatic cell genetic interventions to cure a particular genetic disease, the ethical concerns are the traditional ones: informed consent, risk assessment, cost benefit ratio, sound science, independent review. With germline technological interventions the persons affected do not yet exist, and therefore informed consent is out of the question.[169]

So too are risk/benefit assessments. The ethical question, however, is pressing because of the potential effect on future generations. If germline engineering is used to eliminate certain "unwanted" characteristics, or to insert "improved" characteristics from other species, the human species itself would be "re-created." This would mean passing down for all times human "improvements" which could be deadly, distorting, diminishing, etc.

Cloning is a new technology, and debate about its possibility (or inevitability) can certainly be engaged from within many different ethical

perspectives. The ethical concerns about human cloning are many: the history of hubris, destructive use of new powers, the influence of economics on human cloning, the assault of cloning on what we understand as the structure of a human family and the uniqueness of each human person, the established order of human procreation.[170] Persons are more than their genes, but genes are important. Perhaps more important than genetic uniqueness, however, is the fact that the cloning of Dolly took 227 tries and it is hard to imagine anyone approving the loss of 227 human lives at various stages of development in order to get "the right product."[171]

Human cloning also raises moral concerns related to motives: cloning for genetic enhancement; cloning to produce organs for another person; cloning to replace a lost person. All these motives raise questions about turning one

person into an instrument for another person's intentions. A serious ethics of technology, however, has to be concerned with more than motives. It has to address the ethics of the technological intervention itself. Human cloning in a culture which commodifies everything, certainly threatens the moral status of human life. Just imagine down the line an extensive and widespread industry of human cloning. Business will drive cloning, and that means discrimination will be embedded within it. Human life, rather than a source of wonder, will become another commodity. Human traits, like everything else in such a culture, will be for sale.

As frightening and horrific as a war using nuclear weapons would be, most likely the human species would survive if only marginally. But with the newest genetic technologies applied to biological weapons, deadly microbes could be produced which would out-produce all competition and quickly eliminate human life. Once the genetic codes of viruses and bacteria are more widely available, it is easy to imagine individuals using their private technologies to produce deadly alterations of viruses and bacteria which could not be controlled. Indeed private persons could manipulate the genetic code of microbes to create their own diseases. If nuclear bomb making is available to everyone on the Internet, imagine biological bomb making being made available to everyone.

Even using genetic technology to alter basic foods like corn, wheat and rice raises questions and worries. The risks and dangers of certain alterations of these basic crops may not appear immediately, but by the time they do appear they may be beyond remedy. Some of the opponents of genetically-altered foods may exaggerate the dangers and may over-respond to alterations. But no one can deny that there are risks and dangers associated with genetically altered crops. Consequently there are reasons to begin thinking about where to draw the line and how to limit the employment of this technology.

Genetically altered corn has already proliferated widely. Now it appears that pollen from this genetically engineered product can kill monarch butterfly caterpillars. Not only is this an unforeseen effect of the millions of acres of the genetically altered crop, but the effect has extended to plants in and near the engineered crops. The genetically modified corn produces an insecticide (Bt) in its tissues, and the insecticide is contained even in windblown pollen from the altered plant. This is an unforeseen effect that is still being argued about, but it may provide a metaphor for questions about limits for genetic engineering technologies.

Instead of a killing effect on the butterfly population, a genetically altered crop might have an unforeseen but killing effect on some critical component in the human food chain. The natural structure of farming and food production is delicate, indeed fragile. Genetically altered plants could be so toxic and so reproductively aggressive that the whole food supply could be endangered. The more people have access to the genetic codes of plants, the easier it is to imagine threatening but unexpected consequences from genetically-altered foods. The more extensive the genetic information, the more effective the engineering technologies, the more persons with access to them, the more likely it is that threatening consequences will occur.

The need for an ethics of technology is obvious, but not at all obvious is how to design such an ethics.[172] Scientists and engineers insist on the need for freedom, and they have a point. Yet, ethics cannot be left to the discretion of those actually involved in technology development and employment. We know from extensive experience that they may be motivated by personal economic benefits and other forms of self-promotion. An ethics of technology which can set defensible limits is a major challenge both for humane medicine and liberal Catholic bioethics.[173]

ETHICS OF TECHNOLOGY IN HISTORY

a) The Classical Greek Vision

In the Nicomachaen Ethics, Aristotle gave a start to the quest for an ethics of technology. The right or the good for Aristotle derives from what reason demands in its confrontation with reality. His formula for an ethics of technology is expressed in the phrase *recta ratio factibilium*: what reason requires with regard to those things which we are able to do. Reason, he held, can show us not only what human beings are able to do, but also ethical limits in the sense of what human beings should not do. Reason can show us our positive and negative obligations toward that reality which surrounds us and with which we interact.[174]

The ancient Greek philosophers made a distinction between reality into which human beings could intervene and reality which was beyond human intervention. Some things, they thought, were a certain way by necessity. *Ananke* is the Greek term which they used to express this idea. *Fatum* (fate) or *necessitas absoluta* (absolute necessity) are the Latin and English translations. Human beings had no control over and could not intervene to change these things. For example, objects fall; the sun rises in the east (they thought) and sets in the west; after spring comes summer, after autumn, winter. These realities, they thought, were beyond human technological intervention. They provided the limits to human decision making. Confronted with these realities, humans had only two options: (1) to bow, to venerate, to adopt a religious attitude, and (2) to try to understand. Understanding of *ananke*, or absolute necessity, stopped with the mind. It did not proceed to human intervention or to attempts to change the reality.[175]

Another reality is not unalterable, but rather is amenable to human intervention. This other reality, the opposite of *ananke*, the Greeks called *tykne* or *techne*. This reality is the way it is, not by necessity but by chance, and therefore could be changed. It is open to human intervention. Here technology enters the picture. *Techne* later comes to mean a system or set of rules for making changes and doing things differently; a means whereby things are made.

But there were yet other subtleties in this early Greek ethics of technology. Besides necessity and chance, there was a reality, which is moderately or relatively necessary. Unlike gravity and season change, which they considered absolutely necessary and unchangeable, there was necessity that might never come to be; for example, the disease a person might have but might not have had. The disease happens, but might not have happened. Later Latin philosophers called this *necessitas conditionata or necesitas ex suppositione*, conditional rather than absolute necessity. Here for Greek and Latin thinkers was another area for technology intervention. Humans could intervene depending upon their understanding, their tools and their courage.

There are different attitudes or postures that humans could adopt in the face of conditioned necessity. Humans could confront this reality with a certain bravura or "in your face" attitude. They could adopt what we might call a technological aggressiveness which wants to test the limits in order to see what can be done. A different possibility would be a hands-off attitude or "let it be" stance. This latter attitude is more reverential toward nature. Some environmentalists today would exemplify it.

The very mention of ethics comes over as a threat to the technologically aggressive with their fearless, interventionist dispositions. At the other extreme there are persons who prefer no intervention; no alterations of nature. They would have us live in the woods like Theodore Kaczynski. For them, any and all technology is suspect of being harmful and therefore wrong.

Most people do not think that either extreme is reasonable. But the extremes pose the ethical questions for us. If there are limits to technology, what are they? Where are they? How do we discover them or create them? When do we leave nature alone? When do we intervene, alter, and use?

Humans are structurally interventionist or technological and cannot change who they are or abandon their way of being in the world. Today, we see most reality as what classical thinkers would call conditional necessity. We think we can change almost anything for the better or alter anything for our advantage. Widespread advocacy of unlimited technological intervention is typical in American culture. But we also find advocacy of reverence toward everything "natural" (e.g., the organic food revolution and committed environmentalists). The challenge is to locate an ethics of technology between one and the other.

If we use Aristotle's structuring of the question, we set out to find what right reason or good sense tells us about what we ought to do and what we ought not to do with reality that is open to human intervention and change. That reality is expanding. Once almost everything was considered the way it is of necessity. Now almost everything is considered fair game for science, engineering and technology.

b) A Christian Version of the Classical View

Besides a Christian ethics based on scripture, there is an ethics founded on an understanding of nature. The Greek word for nature is *physis*. Catholic ethics, like early Greek ethics, is physiological. The foundation of morality is the way things are or the way nature functions. If technology imitates nature, promotes it, permits it to function better, or corrects its dysfunction, then it is ethically right. If it does something other (destroys, subverts, inhibits nature's perceived structure), it is ethically wrong. The right and the good are objective. Morality is founded on reason's understanding the way things are or the way they function physiologically. Catholic thinkers in history held to Aristotle's aphorism: "If one way be better than another, that you may be sure it is nature's way." (Book V, p. 131).

The justification of this ethical theory is religious. The assumption is that God, through evolution, made things the way they are, and that the way God did things should not be altered. Human beings are in a very special way God's creation. They are made in the image of God. How are humans most like God and closest to God? In the soul, which the Greeks and some medieval philosophers equated with the mind. Humans are most like God in their intelligence, in reasoning, in knowing. Human knowing is finite and limited, of course, but still humans are closest to God when they understand their own reality and the broader reality that stands over against them. What intelligence or reason reveals is not dissimilar from God's own understanding. Reason shows us the structure of reality (developmentally) and discovers over time the moral standards embedded in it. St. Thomas Aquinas is most closely identified with this viewpoint.[176]

c) A Radically Different Christian View

A very different ethical standard for judging right and wrong appeared in the 14th century. It came from John Duns Scotus and his disciple, William of Ockham.[177] These two medieval Catholic scholars were different from St. Thomas dispositionally and theologically. Duns Scotus, like St. Thomas, believed that we are all made in the image and likeness of God. He, too, thought we were different from all other divine creations. But for Duns Scotus and William of Ockham, we are most like God in our freedom rather than in our intelligence. We act morally not when we understand and imitate nature but rather when we use our wills and our freedom to dominate it. Scotus and William of Ockham did not deny the importance of reason, but they saw human likeness to God not in intellect or in insight into the way things are, but rather in will and in decision making. Will is what makes something good. Something is good because God wills it, and not vice versa. The will and not the intellect is man's superior power.

Scotus said: "*Intellectus, si est causa voluntatis, est causa subserviens voluntati.*" Intelligence in effect serves the will and is subordinate to freedom. If it is will and freedom that make humans most like God, then they imitate God and are most God-like when they intervene, alter, fashion or refashion the reality which they confront.

This philosophical perspective changes everything. It generates a totally different ethics. Technological intervention is not limited to imitation of the way things are or the way things function. Human beings interact with reality as with something which awaits formation and reformation. This ethics of technology is an ethics with fewer restrictions. Reality, which cannot be altered, remains as a limit to human intervention. Intelligence and understanding of limits (*Ananke*) continues to play a part in this ethics, but it does not rule.

Instead of technology being justified by imitating the way reality is structured, now it is justified by human domination of reality. Now humans no longer imitate nature, but use their imagination to overpower it. Human beings are right and good and God-like when they take charge of reality. Here is an ethics of technology that will justify the Industrial Revolution and endorse today's biotechnological revolutions. Science and machinery are tools in the hands of human power. With science and math and the instruments of technology, the possibilities of changing the world are limitless. Ethical use of technology no longer depends upon copying what God has done. We see this perspective reflected in the attitudes of Galileo, Copernicus, and Newton, and then in all the major players in the technological revolution.

In the 17th century, Francis Bacon (1561-1626) proclaimed that knowledge, especially scientific knowledge, is power and is ordered toward power.[178] It was to give man power over nature and enable him to transform life on earth. In the 18th and 19th centuries, science and technology generated enormous optimism. God-like human beings, the Enlightenment thinkers assured us, should take advantage of all the forces of nature. Nothing is impossible. Disease and illness, like all other reality, will be conquered. Reason is still important, but reason is at the service of will and freedom. By the 19th century we see this new ethic taken to the extreme by Nietzsche (1844-1900). Nietzche's vision is pure will or the Will to Power. Now will without limits is the standard of morality.[179] Superman alone is really good. He is beyond all established ethical standards of good and evil. Scotus started a development which radically changed the Western perspective on an ethics of technology.

In the Enlightenment Era, however, something of the older vision was retained. The will to power, except for Nietzsche, could not create its own "Natural Laws." Science presumed to grasp and then to provide for technicians and engineers an understanding of the way things are, a grasp of reality. *Ananke*, in the sense of immutable laws which technology could not change, was still taken for granted. Newton's law, the laws of thermo-dynamics, Einstein's law of relativity were examples. Technology had to work within certain conceptual limits.

d) A Modern Ethics of Technology

Twentieth century culture provides us with yet another ethics of technology and with yet another standard for what may be considered a right or wrong intervention into nature. In fields like genetics, ethical use of science and technology neither imitates nor dominates nature. In these fields, imitation and domination have given way to creativity, in the sense of "creating nature anew." Scientists with powerful technologies create what never existed before. Human beings in this perspective are considered God-like, not when they initiate or dominate nature, but when they become co-creators. Humans create not ex nihilo (out of nothing), but from a nature susceptible to creative manipulation. The former ethical limits of technological intervention (necessity and unchangeable laws) no longer work. For today's biotechnicians and geneticists and molecular biologists, the established order has given way to limitless possibility for creativity. The human creations are justified by a philosophical anthropology which sees human beings as most God-like when they create, not when they imitate or dominate nature.[180]

The most startling examples of the new technology come from the life sciences. Human beings have created forms of life which never existed before: new viruses, new bacteria, new forms of animal life. Can we any longer speak of nature's laws as Ananke, imposing limits on scientific and technological capability? If man stands on the moon today, then what he thought were the laws of gravity no longer function. If the older laws of nature are gone, are there any limits to technological creativity? Is everything that is possible also ethical?

DNA can be altered and combined to create creatures that never before existed. Scientists can apply cloning technology to create different beings, even new types of humans. But should we clone human beings or genetically engineer them for people who want more control, or better athletes, or taller children, or smarter ones, or whiter ones? What are the ethical standards for the proper use of these new technologies? Are there any limits to creativity?

If no limits exist, then we have to think seriously, not just about new creations but about monstrous possibilities.[181] We have to think seriously about the possible end of what we have known as the natural world. Science and technology have developed amazing capabilities, and yet we have not developed anything like a corresponding ethical system to address the possibility of misuse and even of disaster.

TODAYS SCIENCE AND TECHNOLOGY

The United States emerged as a nation out of the enlightenment culture. Both historically and intellectually we are an 18th century development. We understand ourselves as a free nation, and freedom is a dominant individual value. Typically for Americans, something is right or good if it is freely chosen. Creativity and power are also important values both in individual lives and in science. An ethics of limits is a very difficult concept for most Americans, especially for American scientists. Autonomy and scientific creativity here are primary values. Limits imposed on science, for research scientists is a tough sell.

Our scientists most often operate with a set of enlightenment assumptions. Science in this view is a process of discovery based upon quantitative observations rationalized via causal mechanisms to prove or disprove a formulated hypothesis. There is no room in this science for human emotion or for addressing philosophical and ethical questions. This view of science represses a broader way of understanding reality which, historically, scientists and philosophers together experienced with a certain awe and reverence. Awe and reverence are what ground a respect for nature and create the starting point of an ethics of limits: an ethics which would encourage science and technology, but restrain technological intervention which might radically alter or destroy the awesome order of reality. It is this awesome, established order of reality which is the source of both science and ethics.

Human beings are unique in their ability to understand the order inherent in reality. Science is one among many ways of human understanding, but in today's culture, the most important way. The greatest of the scientists show that they do more than understand reality in a cold rationalistic way. They became personally immersed in the reality with which they work. They exhibit in their work what could be described as an indwelling in that reality. Their relationship with cosmology or biology could in some instances even be talked about as reverential, perhaps even mystical.

This dimension of science tends to be lost for many of today's scientists. They seem incapable of moving beyond the more banal tasks of observing and measuring narrow aspects of the whole. Counters and gatherers of data consider themselves scientists in the full sense of the term. Indeed, they hold themselves up as models of what science means, and yet they either ignore or have repressed an important aspect of their discipline: the awe and the indwelling experienced by great scientific geniuses. In addition, they ignore the important distinction made by Francis Bacon between fact gatherers and real scientists.[182]

The great scientific geniuses were truly creative. They were tuned into the inner rationality and established order of the universe. They dwelled within that broad established order. They identified personally with what they were trying to understand. Their personal attitudes and intuitions, feelings and aesthetic sensitivities were as much involved in their science as counting and fact gathering in order to prove or disprove hypotheses.[183] Their science involved their whole person and not just their capacity for cold objectification.[184]

THE NEED FOR SCIENTIFIC AND TECHNOLOGICAL LIMITS

Many of the intelligentsia, just one generation ago, were convinced that Communism would eliminate human poverty and penicillin would eliminate infectious disease. We know what happened with Communism. It extended the poverty it claimed to eliminate. And instead of antibiotics eliminating infectious disease, we are now on the edge of a pandemic created by antibiotic misuse. Multi-drug-resistant microbes were created but never anticipated. These facts force us to think about untoward consequences, the possibility of disaster, and therefore about the need to establish ethical limits to technological interventions into reality.

Scientists formed in a scientific perspective which emphasizes only will power and creativity and having only anorexic counting capabilities are more likely to interact with reality in a destructive way. Since they have no personal regard for what they are trying to understand, they are more likely to intervene into nature with hammers and clubs. Then, they may presume to reorder or reorganize the pieces as they "will," or according to their "creative instincts." They can remain narrowly focused because they have no personal sensitivity for the broader order of things. They can do brazen things with reality because they do not care about it personally. They may be more interested in individual fame and money than anything else. The experience of awe which puts the importance of an individual in perspective has been lost. The coherence and rationality of the larger reality escape their notice and concern. There is nothing about that broader cosmic reality which could cause such an individual to modify personal ambitions or moderate the use of his or her technological instruments.[185]

And yet science and technology, like every other human endeavor, has to be bound by ethical constraints. Scientists and engineers cannot do anything they like any more than popes or presidents can. Human activity always has a moral dimension and must be subject to moral constraints. Evil is possible. Ethics attempts to identify the evil and to establish constraints. It neither undermines science nor unnecessarily restrains technology. Required courses in philosophy of science and ethics of science are appropriate in every graduate program in science.

A scientist must be free and must be allowed to be creative, but the limits to freedom and creativity derive from respect for the broad rational order of reality itself. The great scientists showed this respect and acted responsibly. It is the scientists at the mediocre level and below who are the worries. They now have access to powerful technological instruments. Seemingly small changes or small technological interventions can now cascade into enormous disruptions which could destroy the very conditions for life. It is human life that is especially vulnerable. In a conflict with bacteria or viruses or fungi, it is the human beings who most likely will not survive.

People today have become used to scientific breakthroughs, and most often they expect the results to be positive. They accept breakthroughs like the sequencing of the human genetic code. They remain hopefully accepting of the therapeutic technologies generated by the new genetic science. That hope has not been much diminished by the tragic failures of so many attempts to turn a better understanding of the genetic code into effective therapies for genetic diseases, disorders, or deficiencies. Most people, however, have not even considered, let alone taken seriously, the possibility of genetic technologies not just killing the subject of the experiment[186] but threatening the whole of human life. Diseases and disorders can now be introduced into the genetic structure, which can replicate and then spread beyond control. A genetically altered reality could leave no place for human life.[187] A vector used to carry a gene to cure a disease or to enhance a characteristic could wind up being the only surviving form of life.[188]

Genetic science and technology are not the only threats. Any number of new technologies generated by contemporary science and engineering can create instruments of mass destruction. And many of the powerful new technologies can be accessible to groups and individuals, among them the sickest, the most hateful and the most destructive. Grave concern about just this possibility has been expressed by some of our most respected contemporary technological geniuses.[189] Frightful evil is not just what rogue governments can do. Now even ordinary laboratory scientists and technicians can cause widespread destruction.

THE FOUNDATION FOR AN ETHIC OF LIMITS

Given this new situation, how can the obviously necessary ethical limits to technological interventions into reality be established? How can scientific and technological education be re-connected with ethical responsibility? At the very least, at the beginning of an ethics of technology there should be a critique of certain widespread assumptions, i.e., purely mechanistic concept of science and a notion of creativity that covers anything and everything.

Not to start with such a critique would leave us with a science bereft of moral responsibility and doing whatever an individual technician sees as potential advancement of a career or an economic bonanza. Too many horrors and terrors have already taken place during the past century for anyone to continue to think that science and technology need not be concerned about ethics. The innocent subjects killed in experiments during the Fascist regimes in Germany and Japan were only a fraction of the millions of innocent persons killed in the 20th century with the aid of advanced science and technology.

The assumption that potential applications of science in technology are inherently good and productive only of good effects is nevertheless still widely held. For true believers, the only acceptable ethics of science is sound science itself. The idea that people can be scientifically and technologically sophisticated and yet morally depraved is ignored. And the ethics that is needed cannot be just an expression of what each person feels. The moral standards and personal responsibility being argued for here do not separate science and ethics, but derive both science and ethics from a broad understanding of objective reality.

An ethics of science and technological intervention into reality is something not just for academic types to struggle with. It is an issue with which scientists, politicians, lawmakers, religious leaders, physicians and hospital administrators, have to struggle as well. Ethics is not something to be played with or thought about after we come home from work in the laboratory or the hospital or the university. The idea of science and technology split off from ethics is not realistic. It is not the way human beings experience reality. It is not the way science works, and it cannot be the way an ethics of science and technology is constructed.

Human beings experience values in their contact with reality. Ethics and values are not just subjective feelings disconnected from objective experience. They are part of the push and shove of our contact with reality. The idea that facts are public and universal and ethics is private and individual cannot survive critical examination and cannot be the assumption from which an ethics of science and technology is constructed. Ethics was split from fact and focus on fact by enlightenment and modernist thinkers. Post modernism, in reaction, turned everything into subjective value (how I see things) and private feeling (what I feel is right). Both perspectives are seriously flawed. Neither can generate the ethics of technology which we need to survive.[190]

If biological and medical science and technology have to be subject to moral constraints, where is the foundation for these constraints? What is the ground from which moral limitations can be constructed? What is the justification for establishing limits? In the classical Greek philosophical perspective, it was the inherent order of objective reality (Anake). If updated and modernized, this perspective could still be a defensible foundation for an ethics of technology. To do what is right with new technological capacities has to begin with respecting the inherent order of objective reality. The foundation of such an ethics of technology comes not from outside science but from within science itself. It is reality itself, broadly understood and appreciated by science, which can set up the standards for an ethics of technology. What we come to know from a broad science gives us insight into what we ought to do and ought not to do technologically. Our intelligent contact with an understandable reality provides the foundation for an ethics of responsibility.[191]

Any defensible ethics of limits has to be based on objective reality, on truth which is more than just "what I think or feel." A defensible ethics of technology for today has to be derived from a different way of knowing objective reality. The challenge is to know that reality in a way that makes possible both objective science and ethical responsibility in applying science and technology. Bad science in the sense of science based on a false split between objective fact and ethical responsibility is dangerous. It can generate technologies which destroy both human life and the whole of reality. One way of talking about Natural Law is to talk about respect for the established order.

Good science that reveals reality and responsible technology protects the inherent order of objective reality. That reality hides clues which good science unveils and values which an intelligent ethics articulates. That reality may yet reveal itself in different ways to future scientists and ethicists, but we are hopeful that the future revelations will have some continuity with today's science and ethics. A reference to future scientific revelations assumes that we leave in place an established order.

THE BEGINNING OF A DEFENSIBLE ETHICS

The first period of modern science and technology brought great advances to humankind. Over the years, however, ominous negative consequences have made their appearance. For the good consequences to continue and the evils to be recognized and constrained, a convincing ethics of technology is needed. The argument being made here is that the first step toward constructing such an ethics is found in classic philosophy and medieval theology. In today's complicated circumstance, that older conceptual tool needs to be broadened, more subtly articulated, and more carefully applied. That first step is identification of a basic ethical principle, but it is not all that is needed for a final and defensible ethics of limits. Every real ethics is articulated from a perspective or a vision.

We cannot be satisfied with and simply copy the medieval or classical Greek vision. But working on refinements of their vision and their concept of respect for an already ordered reality is more than reasonable. An updated version of that basic vision can provide scientists or lawmakers alike with a perspective from which technology can be encouraged but limits can be considered. It can provide a justification for constraints when the awesome order of reality itself is threatened or when the subtle interrelationships of background biosystems are endangered. This broad, background vision may not be shared by everyone, but enough people can appreciate its worth to justify using it to begin deliberating about limitations in a democratic society when concerns are raised about survival of the ordered objective reality.

Respect for the intelligible structure of created reality provides a perspective or a place from which to begin the process of judging particular technological interventions. It is not the end of the process. It is the most abstract level of ethical discourse. The facts and circumstances of specific projects or particular interventions have to be thoroughly understood and then assigned an appropriate ethical weight. This is a more concrete level of ethical discourse. Past experience, enshrined in already articulated policies, codes and laws, also has to be considered. Finally, as was mentioned above, some recognition of a place for creativity and will is always appropriate. Adoption of a primary vision or perspective does not mean that conclusions about right or wrong can immediately be assigned to a particular proposal. Adoption of a primary perspective is the beginning of a process of ethical discrimination. It is not a machine which spits out quick final answers about concrete questions. The perspective or vision being suggested here is meant to provide direction for a process of reflection within a liberal perspective on the Natural Law tradition.

This most abstract-level ethical directive is not anti-scientific and not anti-technological. It assumes that human beings can understand the structures of reality. It supports the value of searching for better understandings of reality in order to bring about human progress. But it demands that lines be drawn and limits be established whenever a technological intervention moves toward destroying reality or refashioning the established order. No matter how great the power developed in science and engineering, the established order of things cannot be cast aside or radically altered.[192] Medical technology, for example, can intervene into human bodies but not to the point of reducing the human being to an object. The power to make human beings into things and products instead of procreations is with us in 21st century (reproductive cloning), but the ethics being suggested here would say no. Human creativity and will power at some point should be required to submit to the prime directive coming from a reigning ethical vision. That directive is: respect for the established order in nature.[193]

Quick, automatic, negative answers to questions about human cloning are not responsible. But if the cloning of human beings can be shown radically to alter or threaten to destroy the structure of human being, or human family, or human community, or the sanctity of human life, then it becomes unethical. For example, technologically engineered beings for performance of slave-like tasks would be wrong because it would destroy essential characteristics and the established order of human life: dignity, equality, respect. Life can be relieved of diseases and disorders, and that is good. It may even be enhanced in certain ways. But its personal structure cannot be destroyed.[194]

In December 2000, a 19-page document reported on a meeting between bishops and scientists. The meeting was held in September 1998 on the issue of cloning. By cloning, the bishops and scientists meant the replication of genes, cells, tissues and whole organisms. The bishops and scientists agreed that there is no moral objection to the cloning of animals, or human cells or tissues or genes. The scientists supported embryonic cloning for research purposes. The bishops objected to embryonic cloning because they attributed human status to the embryo and objected to harming or destroying embryos.

Both groups agreed that cloning of human beings was possible and would soon be a reality despite its being morally controversial. The technique for accomplishing whole organism cloning is called somatic cell nuclear transfer. The somatic cell nucleus is transferred to an egg whose nucleus has been removed. The process produces an organism which possesses all the properties of the contributor of the somatic cell. One of the major moral issues is the high percentage of defects in cloned humans/animals; i.e., the extensiveness of harm done. This alone makes the use of cloning for human production now not just immoral but grossly inhumane and scientifically irresponsible.

Bishops and scientists agreed that scientific research is important and must not be discouraged. Human life, too, it was agreed, needs to be defended. But the scientists did not agree that human life from the moment of conception must be defended, certainly not the same as fully formed human life. The two groups also disagreed about two issues: that the embryo was the subject of human rights, and that reproduction must not be separated from marriage. There was agreement, however, about concern for social justice, the role of commercial interests in the development of cloning, the inadequacy of today's cloning science to make the leap to becoming a procreative option and consequently to alter the structure of human life. Not every issue was settled but there was extensive ethical agreement achieved.

Related to cloning but closer to public decision making is the issue of stem-cell research. The liberal/conservative split shows up clearly on the stem-cell questions. The conservative view is articulated by the U.S. Conference of Catholic Bishops which in turn reflects a Vatican perspective. For the Bishops, stem-cell research is a violation of the sanctity of life. It is comparable to abortion. The use of embryonic stem cells is seen as a use of human life as a means to an end and as such the violation of a categorical imperative. They emphasize that no one yet has ever been helped by embryonic stem cells. They insist that adult stem cells are a morally non-controversial alternative. The Bishops equate an embryo with a human person and consider the use of embryonic stem cells equivalent to the killing of a person. This official Catholic teaching stands in opposition to official positions taken by governments in France, Canada, Great Britain, and Germany and comes over as clear but extreme.

Liberal Catholics are not so quick with definitive negative judgments. They tend to agree with scientists that studying stem cells will reveal important insights into the basic biology of human beings and will lead to breakthrough treatments for devastating human diseases. They distinguish *totipotent* stem cells in an embryo until four days old (these can develop into all the cells of the body); *pluripotent* stem cells or embryonic stem cells which begin forming after four days and continue for 8 months, (these can develop into most of the cells in the body); *mulitpotent* stem cells which can develop into many body cells but have significant limitations (these exist throughout human life). Liberal ethicists and scientists generally do not equate human embryos with human persons. For them, development counts ethically.[195]

The promise of embryonic stem cells is seen as justification for continuing federal support for this type of research. The fact that so many frozen embryos are available (left over from in vitro-fertilization) which otherwise will be discarded is an important dimension of the reality being morally evaluated. The fact is that thousands of embryos are left over from interventions to overcome infertility and consequently are doomed to become incapable of developing, and the fact that curing some diseases from the use of embryonic stem cells is likely are all important moral considerations. In a liberal view, the human person develops, and the form of human personhood as well as uniquely human capacities contribute to moral status. The human embryo is human life, but the idea that it has the status of a full person is counter intuitive. No one would intuitively treat a microscopic speck of human embryo as a person. From a liberal Catholic perspective, the human embryo has moral status and demands respect. The level of respect increases with the development of cells into human form and human capabilities.

The use of embryonic stem cells for research is not a disruption or an undermining of the created order. Talk about this type research as contributing to a culture of death is rhetorical and not rationally defensible. If this science and technology continues to advance and turns out to have unforeseen consequences that disrupt created order, then a negative moral judgment would be more rationally grounded. In the meantime, a Natural Law perspective supports the careful continuation of genetic research. The scientists should be sensitive to the moral value of the embryo and the moral dimensions of their research. The bioethicists and religious moralists in turn should stay informed about research directions and the consequences of this type science and technology.

The core moral, philosophical and religious question has to do with what constitutes a human being and the need to protect human life. Since it is the core ethical question, liberal Catholics resist solving the problem or answering the question by authoritative Church pronouncements. Natural Law ethics insists on keeping the question open. It requires responsible science, careful moral analysis of the reality under consultation, consideration of consequences, respect for human creativity and human capacity of understanding, and respect for the inevitable technological interventionism of human beings.

JUSTIFYING LIMITS

Plants and fruits and foods of all types have for a long while been tinkered with and genetically altered with simple techniques.[196] Now, however, our technological capability has opened up the possibility of so altering plants and fruits and foods as to cause a plague which could quickly destroy human life. Recognizing that possibility must signal a reconsideration of where our technological interventions are leading. Some alterations could relieve the world of hunger. Other alterations, however, could relieve the world of human life. This kind of good and evil possibility poses the serious question we are trying to address: how do we pursue the good and avoid the evil? How do we support technological advances and avoid technological disaster? How do we support science and technology, but also establish limits to what modern science and technology can do responsibly?

It makes sense, I believe, to give priority to an ethical perspective that requires as the starting point and the reigning value, a respect for the established order. Even using this starting point, however, it is altogether possible that disastrous consequences would follow from the employment of a particular technology. This vision or perspective does not eliminate a role for human will and creativity. This primary directive, however, does tilt interventions away from radical alterations of the established order just at the time such alterations become possible. Its use would help to develop policies and practices consistent with a basic vision. It would not guarantee that disaster does not occur but would orient ethical, judicial and religious decision making away from disaster. And yet, this basic foundation or primary vision from which to begin ethical evaluation of technology cannot simply be taken for granted. It must be supported by convincing argument coming from experts in science and technology, philosophy, law, medicine and religion. I am arguing here for more thoughtful consideration and continuing dialogue among scientists and ethicists.

Whether a culture is religious or secular, justification is required if limits are placed on liberty. Religious cultures root such limits in the concept of a "created" order which serves as the ultimate ground of morality. The world as a created order is a good in itself, not the ultimate good, but nevertheless a real and objective good. Despite its fallenness, the world in a Judeo-Christian perspective retains an order and a goodness; it displays a harmony, a beauty, an intelligibility that have to be respected, and it provides a perspective from which ethical reflection can begin. Interventions and alterations can be justified as long as the established order remains intact. All basic ethical principles (freedom, justice, love, wisdom, dignity, etc.) are grounded upon a created order in Catholic moral theology.

Secular cultures have their own foundations for justifying limits to freedom. Negative liberty is understood as liberty from interference; positive liberty, a freedom to be one's own master and to do what one wishes. The latter liberty is more dangerous because of its potential for terrible consequences for others. Isaiah Berlin said that "it is at times justifiable to coerce men in the name of some goal (e.g., justice or public health) which they would, if they were more enlightened themselves, pursue but do not because they are blind or ignorant or corrupt."[197] Berlin recognized the need for moral limits, especially to positive freedom.

Berlin supported limits on the basis of broad moral standards like justice and what today we call the common good, e.g., public health. He did not argue from a religious perspective and yet recognized that without constraints on the liberty to do whatever one wants, the world and human life are doomed; victims of blindness, ignorance, corruption. If the limits which he proposed presume to restrict scientific creativity and scientific freedom, they require a substantial grounding. He argues that maintaining life and the world is such a grounding, an objective and rational good. His objective and rational good is similar to what Catholic moralists mean by Natural Law and what other religious ethicists refer to as the created order.

Religious and secular thinkers alike refer to the world's order as the moral foundation of limits to what humans can do with their technologies. To use today's powerful scientific technologies simply to do whatever some scientist or technological hacker wants to do is blatantly irrational. The idea of human liberty to destroy human life makes no sense at all. Liberty makes sense only within a continuing world and with continuing life. Liberty to destroy the world and human life is not liberty but absurdity. Therefore the established order of human life balanced with other forms of life in a sustaining world is the background moral foundation from which to examine particular judgments about specific technological interventions. This is so both in religious and secular ethical perspectives.

Whatever puts the world and human life at risk is rationally absurd and morally unacceptable. It is so for Judeo-Christians, for whom both are created by God, but also for secular persons who are rational and responsible. Both believers and non-believers have to exercise responsibility and avoid absurdity. They have to have a basic respect for the world in which we find ourselves and for the dignity of human life. Something like a respect for the established order has to operate in both types of meaning systems. Both groups have to unite against any use of technology which threatens that order. It is too easy today to slip into total devastation by taking little steps toward a precipice.

Nature is not perfect. It is not God. The established order is not always a pretty picture. There are awesome landscapes and sunsets, beautiful flowers, and children who were begotten, not made. But a considerable portion of that established order harasses and disfigures and destroys. These ugly and destructive dimensions justify continuing human technological interventions to improve things. There is an inherent aesthetic beauty even in the smallest elements of matter and life, but there are repulsive and abhorrent elements as well: the pollutants, cancers, contaminated lungs and cerebral palsies. Love and friendship and civility are beautiful, but hate and revenge and self-destruction are also a part of reality and they are ugly.

But the beauty, the order, the majesty of nature deserves to be protected and preserved. Human life is not the only respectable form of life, but it has priority in a religious perspective because the very nature of God is uniquely revealed in human beings who, like God, can know and will and create. For Christians, the human Jesus reveals God's being in a special way. Will, freedom, creativity are activities of the highest order but nevertheless have to defer to the established order.

Respecting the established order first requires its realistic assessment. Human beings can and should use technology to advance the beautiful and the good and to reduce the evil and the ugly. In doing so, however, they may presume to restructure reality. If and when they do, they are more likely to destroy everything.

When everything is dirt and microbes, it will be too late. We need a widely accepted ethical basis for limiting technological expressions of human freedom and creativity. Respect for the established order is as close as secular and religious thinkers alike can come to an agreed-upon moral absolute, one from which ethical reflection can proceed and particular decisions about technological interventions can be made.

This perspective will be opposed by scientists who are informed by a different set of assumptions. Mechanistic, reductionistic, materialistic assumptions are so pervasive in contemporary culture that many scientists and researchers and physicians are not even aware of holding these assumptions or of the mind-set which they bring to their work. Their assumptions are simply taken for the truth, not unlike fundamentalist believers who take their reading of a biblical text for the truth. Many intellectuals will be immediately uncomfortable with the idea of ethical limits. And the idea of potential disaster will strike them as unfounded worry. They may be right, but consider the following:

In a mainline scientific journal in 1994 Marvin Minski from M.I.T. wrote an article about the future of artificial intelligence. The title of his article was a question: "Will Robots Inherit the Earth?" Minski's answer was, "Yes, as we engineer replacement bodies and brains using nanotechnology. We will then live longer, possess greater wisdom, and enjoy the capabilities as yet unimagined." The idea of using technology to alter the structure of human life is perfectly acceptable to him. According to Minski, humans are now conceived by chance. Their nature can be changed. He sees nothing wrong with the idea of humans being engineered. He puts aside both ethical considerations and worries about ominous possibilities. He focuses solely on the value of scientific progress.[198]

CONCLUSION

When the established order of life is threatened, the technology which does so is morally unacceptable. Evil of this magnitude overrides any argument based on creative liberty or scientific progress. Such a use of liberty violates the most basic moral standard. Technology can intervene into the established order of vegetation, plant life, farming, the human body and the human mind, but not when the whole order is threatened. Such use of technology crosses the moral line.

Medical scientists and medical technologies illustrate very clearly the point about moral limits and crossing a line. In hospitals today, life-supporting technologies continue to be used when a patient is dying from a degenerative incurable illness, and the patient is dysfunctional and debilitated, and the dying is imminent and irreversible. Such uses of technology are wrong. Technology that intervenes in order to cure and to support quality life is good. Technology used to prolong death is bad. The line may not be clear to all; but there is a line, and stepping over that line is wrong. There is an order in human life, and death and a dying process are part of that order. Violating that order or not respecting it crosses the moral line.

The example which most clearly illustrates the ethical role of the established order is the sextuplet case. Fertilization technologies are good when they cure infertility in married couples, but they are bad when they violate or ignore the inherent order of human procreation.[199] A married couple cannot raise seven infants at the same time. It is an objective order, accessible to reasonable people, which serves as the foundation and the ultimate justification for saying no at some point. But at some point, technology cannot just ignore the established order. It can destroy it. What the Iowa couple with septuplets saw as a miracle from the perspective of their fundamentalist Protestant faith-bound ethics is a flagrant ethical violation seen from a reason-based Catholic Natural Law vision of an established order.

Cloning intelligent human beings to get stronger people who work harder obviously destroys the order and inherent dignity of human life. Replacing the established order with a cloning or a genetic engineering scheme which violates this most basic ethical foundation is wrong. In the case of eugenic engineering, it is an affront to the millions of innocent lives who have been sacrificed in the 20th century to this idea of improving the human species. To step over that line again is to step on the innocent bodies of Holocaust victims.

Immediately after news of the successful cloning of a sheep by Ian Wilmut and Keith Campbell at the Roslin Institute in Edinburgh, Scotland, Richard Seed, a Chicago physicist, announced his intention to clone human beings. It took no time at all to leap over a reasonable limit (the human line) with no evidence that the person planning the leap had any sensitivity for the negative possibilities of what he was proposing. Perhaps even more frightening was the scarcity of criticism or any serious movement to slow down Dr. Seed's project. Even in such a situation, many scientists objected to any talk of limits to what science and technology make possible.

Some of the physicists who created the atomic bomb years after the first explosion and in the middle of a nuclear arms race which threatened mass destruction of life finally admitted that "physicists have known sin." They finally came to recognize that their technology had caused enormous evil and little good. It had threatened the very survival of human life. But their admission has had little effect on later scientists and the development of newer destructive technologies.

Trying to work through the complicated task of establishing limits to technology will be difficult, but not to work at the task is irresponsibility in the extreme.

XIV: Advanced Alzheimer's Disease: Stopping Nutrition and Hydration Technologies

INTRODUCTION

Currently, Americans are in the midst of a very serious debate about how to use medical technologies in death and dying situations. We are trying to decide how to help people to die. We approach the issue as if it were a new problem. And we are trying to solve problems associated with the experience of dying by legalizing suicide and euthanasia. Those familiar with the debate know that it is centered on rare and tragic cases. In the face of particularly difficult dying situations, the typical American asks, "Whose life is it anyway?" The answer to that question is obvious and so the debate about how to handle death and dying in America is focused on what the individual person wants. We act as if freedom in the sense of individual autonomy were the only value. Advocates for change in the law claim a right to die in the sense of an individual legal right to be assisted in suicide or euthanasia.

The process of making new laws about dying does not respect our best democratic values. Rarely do the poor, the undereducated, the politically marginated citizens have a chance to voice their concerns about proposed legislation. They do not have access to the legislators. They cannot hire lobbyists. They do not contribute to political campaigns. But they do worry about being forced into suicide or euthanasia by proposed new legislation. Poor women, the poor elderly, the poor who are sick and alone, poor blacks and Latinos and Asians, these groups do not have a role in the debate, but they have everything to lose if the legal solutions being proposed are adopted: i.e., if we legislate a right to suicide and euthanasia.

Predominantly Catholic cultures, like secular cultures, will be forced to make new laws about help for the dying. But besides the value of individual freedom, what is best for the community will be recognized in Catholic cultures as a major value. They have inherited from history all kinds of wisdom.

History provides sound and reasonable teachings about the meaning of life and death, and these are critical when one faces the problem of how to care for

fellow human beings approaching death. Without a religious tradition, death and dying lose their meaning. "They shoot horses, don't they?" is a line that captures the absence of a religious sense and its tragic consequences.

I want to look back to Catholic moral teachings deeply embedded in Catholic tradition, in order to take advantage of already developed, understandable distinctions for deciding about the use of technologies for the dying and to take advantage of the moral wisdom embedded in those distinctions. Conservative Catholics today and some members of the hierarchy take stands against this historical wisdom and against what the Church traditionally has taught. Liberal Catholics in this area become guardians of tradition.

CATHOLIC HISTORY AND MORAL THEOLOGY

The first big lesson from history is that people have always helped other people to die. Americans tend not to consider this when they create "new" solutions. Our doctors, for example, are well educated, but their education pays little attention to history and no attention to theology. They are trained today to see medicine as interventions to cure disease and to prolong life. Until very recently, they were not trained in the different ways of helping people to die.

Too many patients in America die at the end of a string of aggressive and intrusive interventions, attached to all kinds of technologies, isolated, alone, and all too often with severe pain. Their deaths speak volumes about the need for an ethics of technology. These high-tech deaths fuel the crusade in America to solve death and dying problems by letting patients kill themselves or by demanding that doctors kill them.

Historically, doctors and nurses felt obliged to help dying patients to control the pain and suffering. Doctors and nurses were focused not just on the physical pain, but the emotional, social, spiritual suffering as well. In Catholic cultures, doctors and nurses didn't say to dying patients, "There's nothing more I can do," and walk away. They stayed with patients who were dying. Their place was at the bedside of a dying patient. Their obligation was first to avoid applying futile treatments to a terminal disease, then to reassure the patients that their pain and suffering would be relieved, that they would retain some sense of control, and that they would not be left alone.

What history and Catholic moral theology can contribute to questions about the moral limits for technology is direction about when to discontinue the technologies aimed at curing and when to shift to caring for patients in the sense of providing relief of pain and suffering. Humane physicians using a liberal Catholic bioethics know when and for what reasons life-sustaining technologies may be discontinued or refused, withdrawn or withheld. They know who can make the decisions and whether the decision is right.

DIFFERENT BACKGROUND MORAL THEORIES

In contemporary U.S. culture, the whole moral weight of the discussion about death and dying is placed on one issue alone: who decides? Once I know who is in charge of the decision, the question about objective right and wrong disappears. Individual autonomy is the only moral directive. Americans were not part of the history which made important distinctions and searched for what is objectively right and wrong when helping people to die. In American culture, if the patient is competent to decide, then there is no problem. If not, then the issue is how to determine the proper surrogate.

Historical medical codes are clear about the necessity of avoiding futile and ineffective treatments. And historical Catholic moral teachings provide guidelines for determining when even effective medical treatment can legitimately be refused in order to let a person die. The basic moral theological guideline requires a balancing of benefit and harm. And the moral tradition shows how to do the balancing and how to identify harm.[200]

Spanish moral theologians (Francisco de Victoria, Miguel de Soto, Francisco Suarez) fashioned a simple, sensible and convincing moral guideline: anything which burdens the patient or is harmful (pain, cost, shame, repugnance, inconvenience etc.) can serve as the justification for refusing even life-sustaining technological interventions. Medical professionals can respect patient or surrogate refusals, and withhold or withdraw any technology which the patient or patient's surrogate decides is very burdensome or harmful.

This sensible, reasonable, convincing ethical tradition presupposes that life is a basic good, but not an absolute good. This means that we have to work to preserve life, but everything possible need not and should not be done to preserve life. Both moral and medical history support the tradition of helping people to die, not by killing them but by withdrawing and withholding even life-saving technologies when they are burdensome or harmful. After treatment is stopped, the focus of medical help is on caring, especially on pain relief.

What drives the crusade to make suicide and euthanasia legal in the U.S. is experience with a family member's death in which the traditional wisdom of stopping technological interventions at some point was ignored. Consequently, a dying family member winds up being tortured with technologies which are harmful because they provide more burden than benefit to the dying patient. What drives the crusade to legalize suicide and physician killing in America is (1) fear of "high-tech" dying, (2) overemphasis on patient autonomy, (3) worry that dying will ruin a family economically, and (4) repugnance caused by watching a loved one's dying prolonged.

Nothing in Catholic moral theology could justify continuing to do everything possible to keep a person alive. Life is precious and sacred in Catholic tradition, but we cannot do everything possible to preserve it. Could you imagine being obliged to do everything possible to keep someone alive? Doing everything possible would prolong the dying process and exhaust scarce medical resources and deprive things like vaccines for needy children.

I made this point before, but let me say it again. Historically, doctors always have helped people to die. In Catholic culture, help comes in the form of respect for patient decisions to refuse what is burdensome or harmful or not sufficiently beneficial. Once technologies are withdrawn, then providing care in the form of relief for pain and suffering becomes the medical objective. This is what the tradition of Catholic moral theology teaches us. It remains one of the cornerstones of Catholic bioethics.

May physicians provide relief of pain and suffering even when the pain-relieving drugs may or perhaps will stop cardiopulmonary functions? The Catholic moral tradition responds affirmatively. The act of relieving pain is a good act. The intention is to help and not to harm the patient. The possible bad consequence of stopping cardiopulmonary function is out-weighed by the good effect of pain relief for dying patients. And the pain relief is not provided by the patient's death or by the bad result. This is traditional Catholic moral theology. It is called the "double-effect principle." It is accepted in both civil and ecclesiastical law. It is reflected in the bioethics policies of hospitals and health-care institutions. Why? Because it makes good common sense. It is a reasonable response to life when inevitably life enters a dying process. It is an historical example of defensible limits of technology.

THE NUTRITION AND HYDRATION DEBATE

One issue which Spanish moral theologians solved centuries ago with a reasonable, sensible, convincing approach had to do with nutrition and hydration. And yet the same issue is still a point of controversy for some conservative moralists. The nutrition and hydration issue is like ants which get into a house. Just when you think you have finally gotten rid of them, they are back. Some conservatives, many of whom are part of the Vatican bureaucracy cannot seem to let go of the issue of technological nutrition and hydration. They call it feeding.[201]

Feeding is basic care. Sick people must always be fed. Dying people have to be fed when they are hungry and provided drink if they are thirsty. Universally, in every culture and religious tradition, feeding is identified with care. Starving another person or refusing to provide food and drink is seen everywhere as wrong and inhumane. If a person is dying and wants food or drink, refusing it would be an unpardonable abandonment.

But feeding a dying person in the sense of giving that person food and drink when he or she is hungry or thirsty is not the same as hooking a dying person up to a technology which pumps nutritional materials into veins or inner organs.[202] This medical technology is a world apart from eating and drinking. Like every technology, it may be beneficial to a patient, but it can also do terrible harm. Automatically forcing this technology on persons who do not want it or do not find it beneficial, qualifies as assault and battery.

Even conservative moralists have to admit that forcing nutrition and hydration technology on a dying patient is wrong. (There is no obligation to die with a full stomach.) But their definition of a dying patient applies only to persons at the last moments of life. They also have to admit that sometimes this technology causes harm to a patient and therefore cannot always be required.

Whether or not to attach a patient to nutrition and hydration technology is a question that has to be faced today by patients and families, doctors and nurses, hospitals and legislators. It is a decision that everyone who makes out a living will has to face. Finally, it is a decision which most often troubles the family members of patients with advanced Alzheimer's disease. It may require making a judgment about what is best for a patient without an advance directive or about whether the patient is actually dying.

Determining that a patient is in the dying process and that nothing should be done to prolong the dying is not easy. No hard, scientific standards can be applied in order to know with absolute certainty that a patient is dying. But to say that patients are dying only as they take the last breaths is unreasonable and leads to the assaultive applications of futile medical interventions until the end. This kind of unreasonableness creates inhumane medicine that anyone would try to avoid.

In order to avoid inhumane use of technologies at the end of life, a reasonable, if not perfect, set of standards for identifying the dying process is called for. The following are presented for the reader's consideration: (1) the patient suffers from a specific deadly disease or illness; (2) it is progressive; (3) it is irreversible; (4) dying is imminent/threatening; (5) the patient is debilitated, (weak, bed-ridden); and (6) the patient is unable to perform even basic human functions (walking, dressing, feeding). Usually only a physician makes this judgment. A physician who has experience with dying can usually determine that technological interventions are ineffective and cannot reverse the dying process. A good physician can distinguish the following: life threatening illness, terminal illness but not yet dying, the imminent dying process. A dying process, however, does not require unconsciousness and is not restricted to the last moments of life.

The idea of not forcing nutrition and hydration technologies on dying patients is rooted in centuries old Catholic moral teaching developed by the Spanish theologians from the 15th, 16th and 17th centuries. These Spanish theologians recognized that some means to preserve life which ordinarily are considered easy, in the dying process become too burdensome to be required. Too burdensome for them meant too painful, too costly, too shameful, too far away, too repugnant, too __________ (you fill in the gap).

In one of their examples, if a certain food was needed to preserve a life and was available only in some distant land, the patient would not be obligated to leave family, community, culture in order to obtain it. The same is true, they argued, if that certain food were costly or nauseating or repugnant. These theologians developed the concept of *extraordinary (disproportionate)* treatments which meant that the burdens outweigh the benefits from the patient's point of view. When this is so, any medical intervention can be refused, including foods and certainly including nutritional technologies.

Conservative Catholic ethicists and some career-minded bishops today have rejected or ignored this centuries old Catholic moral tradition. They hold that nutrition and hydration technologies have to be provided as long as the patient is not at the door of death and as long as it does not do physical harm to the patient. For them, withholding or withdrawing this oftentimes intrusive technology is equivalent to killing the patient. According to their reasoning, even patients in a persistent vegetative state have to be maintained on this technology. And every person making out a living will would have to demand artificial nutrition and hydration, no matter what.

Stop and think about this. Can you imagine a large Catholic hospital with hundreds of beds all filled with dying patients attached to this technology? Can you imagine the expense? Can you imagine the burden on patients and their families? Can you imagine the scandal? Any reasonable person looking at such a scene would surely ask: Are these people crazy? Do they believe in life after death?

According to this extreme viewpoint, technological nutrition and hydration is not extraordinary (or disproportionate), even when the patient is vegetative and/or dying. It is beneficial and not burdensome. And I ask, "really"? Can we use reason to defend the proposition that this technology benefits persons who have lost consciousness permanently? Can it benefit persons in a persistent vegetative state? What about persons who cannot experience hunger or thirst? Is continued vegetative existence deprived of human capacities and experience a benefit? To whom? Or is it doing harm to the patient's values, dignity and family? Does it not testify against the belief that there is something beyond death?

THE ADVANCED ALZHEIMER'S PATIENT

What about Alzheimer's patients? Persons with advanced Alzheimer's disease are not in a vegetative state, but may not want to eat, may not be able to swallow, may try to pull out the artificial nutrition lines in their veins and stomach. Should nutrition and hydration technologies be forced on them? Do they benefit as persons when attached by force to this technology? Admittedly, a purely physical effect will be produced by the technology, but does this effect provide a benefit to the person?

The most recent medical evidence answers in the negative. A clinical study done on patients with advanced dementia appeared in the *Journal of the American Medical Association*. The authors looked for benefit from technological feeding of demented patients and found none. They examined whether this technology prevented aspiration pneumonia, prolonged survival, reduced the risk of pressure sores or infections, improved function, or provided palliation. In the words of the authors, "We found no data to suggest that tube feeding improves any of these clinically important outcomes and some data to suggest that it does not. Further, risks are substantial." The physicians who conducted this study recommended that the practice of providing technological nutrition and hydration to severely demented patients should be discouraged on clinical grounds.[203]

If it can be shown that the person experiences little or no benefit but rather is burdened, perhaps even harmed, then according to traditional Catholic moral teaching, the technological nutrition and hydration need not and should not be used. There is no obligation to provide an intervention which may produce a physiological effect but produces no benefit to the patient as a person. And there is no obligation when burden outweighs benefit. This is the Catholic moral theological tradition, and it still makes good sense.[204]

In literally hundreds of legal cases, courts in the U.S. have agreed with the Catholic theological tradition. Even the U.S. Supreme Court (in a decision about Nancy Cruzan, a young woman with brain damage who functioned physically, but not mentally) held that mechanical nutrition and hydration was a medical technology which could be withheld or withdrawn. It is not necessary, the judges agreed, to attach every patient to this technology. American legal wisdom, traditional Catholic moral theology, and contemporary medical science, all come together on this point. Also in agreement with this perspective are the American Medical Association (AMA), the American Neurological Associations (ANA), the President's Commission and the vast majority of clinical bioethicists. The few who dissent, (conservative Catholics and Jews), claim that even though patients are not aware of the benefit of this technology, and the benefit does not show up in clinical studies, the fact that it has an effect on their physiology means that it is a benefit to them.[205]

Mainline Catholic ethicists, on the other hand, argue that the patient himself or herself must benefit and the fact that it has some physiological effect is not enough. The effect produced by the technology provides no hope of recovery from the underlying pathological condition. Its effect rather is likely to be a prolongation of the pathology, a prolongation of minimal biological life, a prolongation of the dying process. None of these effects qualify as medical goals or defensible moral objectives. None of these effects can produce a cure, or provide pain relief, or improve the patient's quality of life. If an intervention can do none of the above, then it is not a defensible medical intervention. Medicine does not do just anything that can have an effect. Medicine's interventions have to benefit the person of the patient; and if they do not, they cannot be justified either medically or morally.

If there is doubt about the dying process or about the irreversibility of the loss of cognitive, conative, and relational capacity, then it makes sense to consider mechanical nutrition and hydration, because this technology may buy some time and perhaps, later on, the patient will move toward recovery. But after a reasonable time, the absence of change, and no recovery, then the technology is no longer defensible.

The withholding or withdrawing of this technology from patients who cannot or do not wish to eat and drink is always a serious decision. The decision has to be made with caution because the patient will surely die. Conservative thinkers claim that it is the withdrawal or withholding of this technology which kills the patient. Liberal Catholic moral opinion holds that there is an underlying pathological condition which is the cause of death. Withholding or withdrawing the nutrition technology is not subverting any real medical goal. If a pathology is present which robs the person of human capacities, and it cannot be reversed, then this technology is not extending life but extending death. The decision to withhold and withdraw this technology from a severely impaired patient definitely shortens the dying process, but the underlying pathology is the main cause of death.

Withholding or withdrawing mechanical nutrition and hydration from an Alzheimer's patient suffering from an irreversible loss of consciousness, awareness, cognition, volition, affect, or relational capacity does not cause a new pathology. It simply stops an intervention which is not benefiting the person of the patient and is not accomplishing a real medical goal. If a pathology is present which robs the person of human capacities, and it cannot be reversed, then the technological intervention is both burdensome and futile. Like the bucket full of holes in the story "Daughters of Danes," the mechanical nutrition and hydration has some effect but never accomplishes the task at hand. One has to keep repeating the procedure over and over because the desired result never comes about. This fits the definition of futility. Since the time of Hippocrates, morally, physicians do not employ futile interventions. And since the 15th and 16th centuries, Catholic moral theologians have justified withholding or withdrawing such burdensome interventions.

In saying that this technology may be refused by families of Alzheimer's patients or by the patients themselves through a living will or durable power of attorney document when they were competent, we are making a rational response to a tragic situation. We are thinking and acting within a respectable Catholic moral theological tradition. If not even minimal recovery of cognitive, conative, affective and relational capacities is possible, then the patient as a person cannot benefit. Without benefit we have only burden, and we can stop a treatment. In fact, we ought to do so.

In Catholic moral theology, when an issue is still being argued, you can follow either an open or a closed stance, a liberal or a conservative view, as long as the option is rational. What you cannot do is use church authority to impose a perspective while it is still under consideration or in debate.

U.S. CATHOLIC BISHOPS ON NUTRITION AND HYDRATION

On April 9, 1992, the Pro-Life Committee of the U.S. Catholic Bishops published a statement entitled "Nutrition and Hydration: Moral and Pastoral Reflections."[206] The statement argues for a presumption in favor of the use of nutrition and hydration in all cases, stating that such means of treatment may be withdrawn "...only if they offer no reasonable hope of sustaining life."[207] Following this statement, many ordinary Catholics and even some Catholic bioethicists assumed that there is a presumption for the use of technological nutrition and hydration and that this presumption is an authoritative stance of the Catholic Church. It is not. A liberal Catholic has two concerns with this assumption: The first relates to the perception of the nature and authority of the document; the second has to do with the substance of the argumentation.

The statement comes from the Pro-Life Committee of the National Conference of Catholic Bishops, not the entire conference of bishops. The Administrative Committee of the conference approved its publication *as a statement of the subcommittee*. It was neither reviewed nor voted upon by the entire body of Catholic bishops.

There is, in fact, a wide diversity of views on the mechanical nutrition/hydration question even among Catholic bishops. The conflicting views of the bishops of Texas[208] and Pennsylvania[209] dramatize this point. To its credit, the committee statement acknowledged this diversity of views, without attempting to close the discussion. The authors describe this as "our first word, not our last word" on this issue.[210] It would, therefore, be a mistake to treat the statement as an authoritative stance of the Catholic bishops, much less the Catholic community in the United States.

UNDERMINING A SOUND TRADITION

The presumption in favor of the use of mechanical nutrition and hydration is presented in the document as an application of Catholic moral teachings. It is not. Instead, it is the application of a statement (taken out of context) from the "Declaration of Helsinki" concerning medical research: a statement which is presented as confirming the insights of Catholic tradition on foregoing life-sustaining medical technologies. This use made of the Helsinki text not only paves the way for an extreme pro-life conclusion, but also represents a distortion of the substance of Catholic moral teaching itself.

The committee statement begins by recalling Catholic teaching about the duty to care for life and health. There is no need to repeat them here. They include directives about the means to be used in preserving life. Drawing upon the 1980 "Declaration on Euthanasia" from the Vatican Congregation for the Doctrine of the Faith, the Bishops' statement invokes the distinction between "ordinary/proportionate" and "extraordinary/disproportionate treatments: "One is not obliged to use either 'extraordinary' or 'disproportionate' means of preserving life, that is, means which are understood as offering no reasonable hope of benefit or as involving excessive burdens."[211]

Consequently, if a technology either offers no reasonable hope of benefit or cannot be used without excessive burden, then it may be judged to be "disproportionate" and may be foregone. In this shorthand way, the Bishops' statement recalls a well-established principle from Catholic moral theology: benefit, burden, and their relative proportionality constitute objective criteria that call for prudent, case-specific application.

The next section of the Bishops' statement, in a question-answer format, applies the general moral principles to questions about technological nutrition and hydration. Question 3 is "What are the benefits of medically-assisted nutrition and hydration?" Then we read: "According to international codes of medical ethics, a physician will see a medical procedure as appropriate 'if in his or her judgment it offers hope of saving life, re-establishing health or alleviating suffering.'"[212]

The statement in quotes is a passage from the World Medical Association's Declaration of Helsinki: Recommendations Guiding Medical Doctors in Biomedical Research Involving Human Subjects. The passage comes from a section of the document entitled: "*Medical Research Combined with Professional Care (Clinical Research).*" It reads, "In the treatment of the sick person, the doctor must be free to use a new diagnostic and therapeutic measure, if in his or her judgment, it offers hope of saving life, re-establishing health or alleviating suffering."[213]

As the title and context of the Helsinki document indicate, the passage refers to biomedical research or clinical experimentation. If, in the doctor's judgment, an experimental treatment offers the hope of saving life, re-establishing health or alleviating suffering, then he or she should be "free to use" such a new treatment. However, the general principles of the Helsinki document make it clear that "free to use" means free to "propose" to use an experimental treatment. "Each potential subject must be adequately informed of the aims, methods, anticipated benefits and potential hazards of the study and the discomfort it may entail. He or she should be informed that he or she is free to abstain from participation…."[214]

The point is that the Helsinki text does not warrant the conclusion that simply because, in the physician's judgment, a treatment "offers hope of saving life" it should be used. In clinical experimentation, the research subjects' assessment of the "anticipated benefits" and "potential hazards" (benefits and burdens) is crucially important. The subject must then give an informed consent. The conclusion of this critical study is clear and unequivocal: i.e., "hope of saving life" is not an obvious benefit that calls for continued use of a technology when the patient is dying or has refused the technology.

This Pro-Life Committee statement misuses the language of the Helsinki text in two ways. First, it totally lays aside the research context of that text. This is important because informed decisions about life-prolonging measures that may be justified in a research context may differ substantially from decisions about what is obligatory in ordinary practice. In the next paragraph, the Pro-Life committee statement goes on to argue that for patients who can assimilate the fluids, technologically delivered nutrition and hydration are beneficial because they "sustain life." The Helsinki statement, however, cannot be used as a source for such a claim. Such a use is either mistaken or dishonest.

The second misuse of the Helsinki text is even more serious. In the concluding sections of the document, addressing the critically important and controversial question about foregoing mechanical nutrition and hydration for patients in a persistent vegetative state (PVS), the Pro-Life Committee judged that there should be a presumption in favor of its use. A liberal Catholic perspective argues just the opposite. What is most objectionable is the significant and unexplained changing of the criteria upon which decisions about the use of technological nutrition and hydration should be based: "Such measures [nutrition and hydration] may be withdrawn if they offer *no reasonable hope of sustaining life* or pose excessive risks or burdens."[215] The same presumption is stated in the conclusion of the document: "We hold for a presumption in favor of providing nutrition and hydration to patients who need it, which presumption would yield in cases where such procedures have *no medically reasonable hope of sustaining life* or pose excessive risks or burdens."[216]

The first criteria for assessing whether a given medical treatment is "extraordinary" or "disproportionate" has been changed from "does it offer reasonable hope of benefit to the patient?" to "does it offer a reasonable hope of sustaining life."? Where did the criteria "sustaining life" come from? The answer is from the out-of-context passage from the "Declaration of Helsinki." The conclusion reached ("the presumption for the use of nutrition and hydration") is not based on the Catholic principles recalled in the first part of the statement, but on the quoted out-of-context passage about what may be appropriate in clinical research.

There is a world of difference between a medical technology offering the possibility of sustaining/continuing a patient's physical life in a clinical context and that same continuation of life constituting a benefit to any patient at any time. This, in fact, is what the discussion is about. Does continuing the life of a patient in a PVS through mechanical nutrition and hydration constitute a benefit to the patient? Put differently, continuing the life of a patient in PVS through nutrition and hydration may be a medical *effect*, but it is not a *benefit*.[217] The statement of the Pro-life Committee is a disappointment because in the most important passages, it asserted what it needed to demonstrate. It changed the traditional Catholic criteria from "hope of benefit" to "hope of sustaining life." Was this done honestly or was it meant to deceive and subvert a respected moral tradition? I don't know.

This is one more example of how some Catholic ethicists can distort respectable and reasonable Catholic moral teachings. They distort in order to make their extreme stand on the life issue appear to be consistent with long-standing Catholic moral teachings about the right to withdraw or withhold even life-saving measures if they are burdensome. The principle of life in the Catholic moral teachings is an important value, but it is not the only value, and it is not an absolute value. And Catholic moral arguments about what reason demands in certain complex situations assumes that those making the arguments are honest.

How does dishonesty if it is present, work out in practice? It causes scandal. The Archbishop of St. Louis in October 2000 used his episcopal authority when he intervened in a clinical case to bar the removal of feeding tube technology for a man in persistent vegetative state. The man had to be removed from the Catholic hospital and died a few days later at home. If the whole hospital were filled with patients in similar conditions, would the Archbishop have stuck to his guns? If so, he would have to be removed from office because his behavior would be patently absurd. Why do such a thing? Catholic moral teachings about technologies in the area of death and dying are based on long experience with dying patients in hospital settings. They provide us with a clear example of rationally defensible limits to technology and of a rationally defensible humane medicine.[218]

XV. Equity and Justice in High Tech Medicine

EXTENT OF THE CHALLENGE

Contemporary medicine is high-tech and consequently enormously expensive. In the context of medicine, equity refers to issues of access to these expensive therapeutic technologies. At the most concrete level, it is expressed in the way patients are treated in the high tech health-care facilities. It is hard to imagine a more complex area in contemporary bioethics. It is hard to imagine a greater challenge to a more humane medicine.

In the wealthiest nation on earth, some elderly people are skipping meals in order to pay for medications and procedures they need to stay alive. Some younger patients with serious illness have to wait until a crisis develops in order to be admitted to a high-tech emergency room, their only access to health care. Some doctors treating rich patients make millions of dollars a year providing enhancement technologies (removing wrinkles and tucking tummies) while other doctors are leaving the profession because they cannot make a living on what they are paid for treating the poor. Some HMOs which restrict critically needed therapeutic technologies are paying out rich dividends to shareholders while other HMOs which provide the same services are in bankruptcy and closing. Medical science is creating new drugs for previously untreatable illnesses, but the drugs are so expensive that most people in the world remain sick and die because neither they nor their health plan can pay for them.

Outside the U.S., in some state-run medical systems which claim to provide universal access, 75% of the population never see a physician. Government supported health care for all citizens began in Russia, but now the Russian health-care system is in virtual collapse. During the Soviet era, things were bad. In the Soviet Union in the early 60's, I saw treatment of patients that was technologically primitive and inhumanely insensitive.[219] Standards of care were shockingly low. Patients were treated with the same indifference that customers experienced in Soviet department stores or banks. Now only the rich in private hospitals get decent treatment. Public hospitals are dirty, do not have even basic technologies, and are overwhelmed with desperately ill people. The health-care system that pioneered equity is now a paradigm of inequity and inhumane medical care.

High-tech scientific medicine today is better than ever equipped to cure and to prevent disease, but most individuals, industries, even governments cannot afford the costs. Wealthy patients may experience the ecstasy of recovery from life-threatening illness, but increasingly greater numbers of economically less fortunate patients experience anger and frustration because of being left to die. (Paradoxically, these economically less fortunate people may at least die quickly and peacefully, while economically affluent people may die only when expensive technologies are no longer able to extend the dying process.) The gap between the wealthy who have access, and the poor who lack access to contemporary high-tech health care is a potentially explosive issue both in countries with a predominantly free-market health-care system and in countries with a predominantly government-run system.[220] Basic ethical components of a doctor/patient relationship cannot be violated without causing serious repercussions.[221]

When richer people or persons from certain races or from certain places have access to the high tech health care and survive, while poorer persons or people from other races or places do not have access and die, intuitively we recognize that basic ethical values are being violated. Equity is violated. Justice is violated. Wherever equity and justice are violated, human beings recognize the immorality, experience anger, and suffer frustration. Then if political movements for changes to remedy the immorality and frustration are unsuccessful, revolution can result. The ethical principles of equality and justice may be abstract and theoretical, but there are potentially serious consequences if they are ignored. Plainly stated, equity requires that essential goods and services which are provided to some persons in a society should be available to others similarly in need. Essential health care should not be available only to some.

If even essential goods and services are so scarce or so expensive that they cannot be provided to all, then according to one theory, they should be made available through a form of lottery.[222] The equal ethical value of each person, it is claimed, would thereby be protected. The logic of such a conceptualization is admirable but providing for essential human needs in an equal way is so complex that the logic does not work out in practice. The logic is simple, but the realities are complex. Equality imposes obligation, but the principle of equality alone, applied through lottery, cannot solve basic distribution problems of expensive high tech therapy. Overly simple schemes to implement equity are no help.

In New York, a small health-care revolution in the 1990's brought badly needed primary-care physicians to newly constructed clinics in the city's poorest neighborhoods. The bold initiative was led by a Primary Care Development Corporation. The system provided what amounted to efficient and affordable health care for poor people who had been taking routine needs to emergency rooms. But the number of uninsured patients increased. Then Medicaid, Medicare, and managed-care payments declined, and primary-care doctors started to move to more affluent neighborhoods. Some clinics finally began to close, and the courageous effort to bring a degree of equity to health care, despite strong political support, is now on the brink of collapse. If this happened in good economic times in the U.S., imagine what it is like now that the economy has slumped.

For equity to work, economics has to work, and many other ethical principles have to be put into practice. Autonomy, for example, cannot be ignored. Neither can sanctity of life. And without compassion, equality could neither determine needed health care nor provide needed services. Equity first has to be clearly defined and then related to economics and to other ethical principles.

UNDERSTANDING EQUITY

Equity in the Natural Law tradition means conformity to accepted standards of justice without prejudice, favoritism, or fraud. Equity and justice are closely related. Justice sets up standards for goods distribution, and equity is one of the standards. Natural Law theory requires that equals be treated equally.

Justice in the sense of distributive justice refers to the allocation of limited goods and services. The distribution of goods and services to everyone on the same basis is one meaning of both justice and equity. Ideally, justice would strive to make all human beings as equal as possible. Justice, fairness, impartiality, equity, these are at the very least comparable categories, different ways of expressing the same Natural Law ideal and objective.

The Natural Law foundations of justice and equality are revealed in human responses to reality. Even small children have a sense of justice and equity. Protest and crying can erupt at a birthday party if one child thinks his piece of cake is smaller than the child's cake next to him. Justice and equity are values which have a foundation in reality and are built into the human psyche, even a small child's psyche.

Sensitive persons at any age can see in the face of the other a basic similarity; similar needs, similar responses to reality, a parity, indeed, a sameness. Equity expresses this intuition into reality in a moral category which communicates both the objective basis of the intuition and the corresponding sense of obligation. Equity connotes a requirement that we try to flesh out the intuition of sameness by trying to bring essential goods and basic services to all on the same basis. If the goods and services are medical, then the obligatory aspect of equity point toward a universal health-care system in which basic medical goods and services are provided to all persons.

A sensitive person can grasp the sameness of all humans, simply by looking attentively into another person's face. But even less sensitive people can grasp this sameness when human beings are ill. Whether illness is physiological or psychological, human beings all experience the same distress, loss, pain, and suffering. Depression is similarly experienced whatever the person's socio-economic status or ethnic identity. Cancer, heart disease, kidney dysfunction, all are experienced the same way by human beings. The needs created by disease and illness are the same. It is the overwhelming sameness in experiencing disease and illness that grounds the intuition that persons suffering from the same condition should be similarly helped or treated the same way. Terminal kidney disease, for example, means that persons either receive dialysis, or a transplant, or they die. It is the same for everyone. This is the objective Natural

Law foundation for the moral claim of equity in health care. But the moral claim is one thing, and measuring concretely what high-tech goods and services the claim of equity requires is something else.[223]

How efficient is a health-care system in providing basic goods and services to all? The answer to this question depends upon how the basic goods and services are identified and measured. The measurement depends upon the instruments of measurement and the background assumptions of those operating the instruments. Children in the U.S., for example, may be considered generally healthy by some measurements, but if data are collected differently or if different conditions are focused on, children here are worse off than children in developing countries. Compared with adults, the death rate among children in the U.S. is low, but compared with children in other places, it is not. Life is too complex for most measurement devices to produce clear, cogent, and unarguable data related to equity in the area of health.

Every society organizes, finances, and delivers health services differently. Health-care organizations attempt to provide this essential human good within the limits created by available resources, competing goods, and reigning political perspectives. Comparing one health-care system with another is difficult because the very definition of health care may differ considerably from one culture to the next. Health care in some cultures like our own may be synonymous with expensive medical technologies for curing particular illnesses. In other cultures, health care may mean prevention rather than curing of illness. Judging equality and inequality cannot be separated from all kinds of background metaphors and socio-cultural beliefs.

Some of the newer medical technologies are more effective than older ones, but no system of health-care delivery could provide the most expensive therapies to all.[224] Who really needs the newer, expensive high tech drugs and procedures? Who could get along with older, less expensive ones? Is making such a distinction defensible? Who decides this? How are the data created? Are the standards fair? Measuring equity is complicated by all these variables.

Almost everyone agrees that primary and preventive care are critically important areas of health care, and equal access to these may seem feasible. In these areas, persons are required to be more responsible for their health: to eat right, to exercise, and to receive routine care. Consequently, the costs are less. But even these less expensive low-tech health-care services are not cost-free. Healthy practices on the part of patients require monitoring. Someone has to monitor blood pressure, lipid levels, sugar levels, etc. Then, professionals have to interpret the data and provide routine care. Even the less expensive primary and preventative care costs something. If economic resources are very scarce, even equal access to these may fail. If high-tech secondary and tertiary care are added to the goods and services, costs skyrocket. Adding high-tech secondary and tertiary care in a system inevitably expands the disparities and reduces equity.

People generally agree that equity is important and should be pursued. But they also have other beliefs. Most North Americans believe in the free market rather than government as provider and distributor of medical goods and services. In other countries, people believe that health care is a government responsibility. Given different beliefs, the variety of delivery systems, the diversity of cultural values, the different economic systems, and the different levels of care, equity becomes a value difficult to measure and difficult to implement.

A Socialist theory of justice measures equity in health care one way, and a Libertarian theory makes the measurements differently. The Socialist perspective leaves out of consideration individual freedom and hard work. The libertarian vision leaves out of consideration influences like genetics and environmental factors. Socialist theory maximizes access. Libertarian theory maximizes personal responsibility. In the Libertarian theory state intervention to concretize equal treatment is considered a violation of personal property and justice. In the socialist view, ambition and hard work is discounted. Libertarianism tends to undermine community and shared benefits. Socialism tends to create inadequate wealth for decent health care.

EQUITY AND THE CONCEPT OF HUMAN RIGHTS

Health insurance was introduced as a way of protecting wage-earning workers who become vulnerable if they got sick. Workers with a basic health-care insurance policy attained a certain degree of equity in health care. Later, governments stepped in to extend basic coverage to other vulnerable groups (elderly and poor). It was the concept of health care as a basic objective human right which made insurance and its extensions possible. Broadly extended health-care insurance gave flesh to the idea of equity in health care as a Natural Law principle and basic right. The human-rights concept provided motivation for industrialists and politicians to implement health-care programs for the needy.

The human rights concept is linked with equity in health care both historically and philosophically. Equity is an old concept, but only in this century has it been proposed as a universal human right. Equity in effect is joined with such basic Natural Law requirements for decent human life as freedom from slavery and torture and arbitrary arrest. It is on the same level as freedom of speech and assembly and religion. Equity in health care is included under the general concept of the right to equal treatment under Law. The inclusion of equality among the most basic human rights certainly puts the continuing campaign for equality in health care on firm ground.

Equality as a universal right is frequently mentioned alongside freedom. Both equity and freedom have roots in Natural Law theory, Roman law, and the Enlightenment. Certain political documents advanced both freedom and equality: the Magna Carta in 1215, the American Constitution in 1787, the French Declaration of the Rights of Man in 1789, the American Bill of Rights in 1791, the UN Declaration of Human Rights in 1948. This last document proclaimed "the equal and unalienable right of all members of the human family to freedom, justice and peace in the world."

The XIV Amendment to the U.S. Constitution stated that no state shall "deprive any person of life, liberty or property without due process of law; nor deny to any person within its jurisdiction the *equal protection of the law*."

The philosophical and political texts express a Natural Law vision, and that vision gradually is translated into concrete cultural and political practices. Remedies for inequality in health care need not wait until everyone agrees on how equality in health care is defined or how it will be measured. Steps toward implementing the value of equality in health care are on going. And implementation is aided by expressing the demand for equity in rights language.

THE RIGHT TO EQUALITY IN HEALTH CARE

Equality as a right translates into the right to equality in health care. As a negative right, it means a right to be protected against serious health hazards. As a positive right it means the right of access to certain basic health-care benefits. Rights to health care begin as moral rights, supported by ethical arguments and a Natural Law vision of humanness. Subsequently, human rights become legalized; i.e., turned into legal rights. The objective of Natural Law declarations of human rights is to express what people need for truly human lives. Legal rights to health-care access attempt to put the moral declaration into concrete practice.[225] Health-care programs try to give concrete form to the ethical vision of equality and the Natural Law based right to health care for all.

The impact of the concept of equality as a universal human right has only begun to be felt. It may take centuries before the concept is concretely established in particular laws in diverse cultures. But the different declarations of equality as a universal right are already having an effect. Changes are taking place here and elsewhere to give flesh to the proclamation of this ideal. Cardinal Bernadin, in Chicago, for example, gave concrete expression to the ideal through his challenge to legislators to make basic health care universal by 2002.

He left the mechanism for achieving equality to the experts in health-care politics and economics. If the politicians and economists meet his challenge, it will be one more example of how a Natural Law vision, and concepts like human rights, and abstract principles like justice and equity can bring about more improvements and better changes than wars and revolutions ever did.

THREATS TO EQUALITY AS A BASIC RIGHT

The trouble is that no matter how great the effort and how much of limited resources is invested in extending health care to all, the ideal has not been realized. Consequently, some have simply given up and substituted autonomy for equity. Individual autonomy joined to free-market capitalism creates a vision that makes health care something which each person pays for from his or her personal wealth. No one, however, is required to pay for anyone else in this vision. If equality in the sense that every person has a right to health protection and health care cannot be attained, then any attempt to approximate the ideal is abandoned. Here is a classic example of the baby being thrown out with the bath water.

Paradoxically, the same concept of rights which once helped propel equity initiatives in health care now challenges the hard won advances. People are concerned about rights, but the concept of rights is now more broadly employed. No longer is there a required grounding in human nature for what is claimed to be right. Rights are not restricted but greatly extended. Besides individual patients claiming a right to access health care, doctors claim a right to decide whom they will treat. Insurance companies and capitalist health-care institutions claim a right to satisfy the financial interests of their stockholders. Industrialists and businessmen claim a right to compete worldwide and not be disadvantaged by having to pay for health-care benefits for workers. Drug companies claim a right to make a profit on the products of their research and therefore to charge exorbitantly for their medications. All these rights claims work against efforts to put into place a natural law-based right to equal access to basic health care for all.[226]

RESPONDING TO THREATS

One way to address the threats to equality would be to downplay the concept of rights and focus on a concept of justice which balances equity and autonomy. Working out the concrete details of a health-care system which balances microallocation with macroallocation, primary care with curative medicine, high tech therapeutic interventions with public health measures, equality with autonomy: this is the challenge.[227]

A rationing system is one way to try to respond to the challenge. Rationing, alone, however, does not accomplish the goal of equity. Health-care costs always exceed patient needs for health-care goods and services. No matter how much rationing is decreed, health benefits for all persons are not provided equally. The rich, the socially well connected, the celebrities, the imaginative, the persistent, the less than honest, always find a way around the rationing no matter how strongly the system tries to promote equity.

Every system of rationing is based on the concept of need. Rationing attempts to meet essential health-care needs of all citizens. But how are "needs" defined? Is any benefit a need? How about benefits which restore normal functioning? Could "need" be correlated with "significant" health benefit? Even if the concept of need is reduced to basic or essential or minimum need, it remains difficult both to define and to meet. What is meant by terms like "basic" or "adequate" or "essential" or "minimum" need in health care?

And there are other needs which make a claim on the same limited resources: food, education, shelter, transportation, police protection, drug prevention, water supply, etc. These are not considered health needs, but certainly they have an impact on health. Resource limits make equality in health care a challenge which may never be perfectly met, but a challenge which can be faced up to, one that can generate creative initiatives, and one that can effect gradual improvements in a health-care system.

COMMUNITY AND COMMON GOOD

A President's Commission in the U.S. recently made a plea for universal access only to an "adequate level" of health care, and did so in terms of community responsibility rather than individual rights. In this perspective, the community, whether local or national, without a health-care system which provides some version of equal access for all community members, is morally deficient. The community-based moral obligation focuses attention precisely on community members who are marginalized and whose health-care needs are not attended to.[228] "Common good" rather than individual rights becomes the foundation of equity and the basis for a community obligation. Common good incorporates the value of equity. Even if the common-good concept does not immediately produce equal access to adequate health care for all, at least it makes possible steps in that direction. And truly human community is grounded on a vision of human nature that is social.[229]

But even the contribution of community and common good to equitable access to adequate health care is not without its limitations and drawbacks. Is the community responsible to provide adequate health care even to those persons who flagrantly ignore their individual responsibilities? And how can individual responsibility be separated out from public pressure created by the advertising of unhealthy products and behaviors like smoking? How can individual responsibility be separated from peer pressure, psychological weakness, genetic pre-disposition, etc? In the U.S. we have a culture of "victims." Intravenous drug users and overeaters and anorexics and alcoholics claim not to be responsible for their health problems. Rather than being removed from community responsibility lists, they claim the right to added community support for health care based on their self-declared victimhood[230]. Besides, the ethics of the medical profession has always required doctors to treat persons in need without judging their responsibility for their problems.

SPECIFIC CHALLENGES TO EQUITY IN HEALTH CARE

We spoke earlier about equity as a more realizable ideal if health care is restricted to primary and preventative care. Is it possible to imagine that a community reaches consensus about primary, preventative and acute health care? If that consensus is reached, what more does essential or basic or adequate care for all cover? What more should all persons have access to: dental care, rehabilitation services for alcohol and drug addiction, nursing home care, pre-natal and post-natal care, family planning services, and supplies? Deciding these questions depends upon resources.

Even in the case where equity is concretized by agreement about primary, preventative, and acute health care for all, poor people usually cannot access all services. Even if they could gain access, the institutions which provide care for the poor are rarely equal to institutions for the rich and well insured. Continuity of care is usually lacking. In effect, even primary, preventative, and acute health care may not be equal. Making basic or adequate or essential care equal and accessible to all is not impossible, but it requires continuing effort.

Take the example of Canada. The fact is that Canada has for 50 years been operating a single party payer system which attempts to provide a version of basic benefits for all. Over those years, continuing efforts, continuing changes, continually increased financial commitments have been made. And yet, Canada today is being forced to face reforms either to reduce the basic benefits or to back away from universal coverage because the cost is unbearable. Basic health-care benefits and services for all is a worthy ideal, but like most other ideals, implementation is a struggle.

Equity must, however, remain a moral objective which drives efforts for a more humane medicine.[231] And effective change starts with a careful assessment of each person's surrounding reality: *Yo soy Yo y mis circunstancias* (I am myself and my circumstance)[232]. Let me mention just a few aspects of that circumstance which create challenges to equity in humane health care and have to be considered.

1. Maintaining basic universal coverage in the face of steady increases in immigrant populations, some of whom migrate just for health reasons.
2. The problem of administrative costs which can quickly consume the resources assigned for care.
3. Micromanagement of physician decisions seen as a necessity for managers and as an intrusion by physicians.
4. Astronomical malpractice payouts, and in reaction, wasteful defensive medical practices.
5. The restraint of raising health-care costs.
6. Co-payment requirements which can destroy equality.
7. The handling of high-risk patients.
8. Effective monetary constraints on medical suppliers and pharmaceuticals.

9. Managing the expansion of mental disease categories and payment for mental health care without downplaying the importance of care for the mentally ill.
10. Managing fraud and abuse which costs as much as $100 billion a year in the U.S.

A FINAL CHALLENGE: DOMINANT METAPHORS

The Natural Law principle of equity (and the different values which it enshrines) struggles for recognition because it lacks a prominent place in our way of understanding medicine. As mentioned above, the prominent background metaphor in modern medicine remains that of war, and the war metaphor has a deleterious effect on medical priorities and medical practice. We mentioned above that when we talk of a war on disease, we mean unlimited war rather than a just war perspective. We *battle* disease, *plan attack*, order *batteries* of tests, search for *magic bullets*. Doctors think of themselves as on the *firing line*, and in the *trenches*. They treat *aggressively*, especially *invasive* cells. They take *heroic action* and use the *body's defenses* to *conduct the fight*. The top doctor in the U.S. is the *Surgeon General*. Some doctors don't talk to other staff members and do not permit input from patients because that's the way tough generals behave during a war.

This background metaphor affects the way medicine is understood and practiced. If death is the ultimate enemy, we can understand overtreatment of the terminally ill and opposition to withdrawing expensive high-tech procedures even at the end of life. The war metaphor joined with high-tech medicine explains the priority of tertiary over primary or secondary care, of critical care over chronic care, of intensive care over hospice care. The point of all this is that the war metaphor has no place for *equality*. Consequently, *equity* has to struggle to find a place or a justification in modern practice. But only when it finds a place in the overall medical system and in the doctor/patient relationship will medicine be sufficiently caring and more humane.

In competition today with the metaphor of war is that of industry. Health care and medical treatment now is *an industry*. Doctors are *providers of services*, and patients are *consumers*. Patient care is *managed*. Concerns are expressed about *productivity* and *cost effectiveness*. Equity may find a place more easily within this metaphor, but it will not be easy. Neither in a war nor in a free market metaphor is there much talk about equity.

A more humane medicine is an objectively based and solid moral ideal.[233] Ideals do not translate easily into concrete improvements in the way doctors treat patients. But they do have an influence. Reforms and improvements take a long time. Based on long experience, liberal Catholics know about being persistent and being patient. A more humane medicine will not be realized tomorrow, but improvements can be made and people can speak out against violations of basic ethical principles. If the speaking out is intelligent and convincing, it has a good chance ultimately of making a difference.

CONCLUSION

The practice of medicine is always the expression of a background vision which may not always be recognized. What is true of medicine is also true of ethics. After outlining the characteristics of a more humane medicine, I tried to explain and then justify a background ethical vision in light of which decisions can be made which contribute to an ever more humane medicine. The background ethical vision is Natural Law theory applied within a liberal Catholic tradition. This book is a refection on bioethics in a broad sense and on the background theory employed to support a more humane medicine. As incarnate beings, we human beings are destined to be ill and to need medical help. A more humane medicine will benefit everyone. Medicine turned into just another business will mean that inevitable pain and suffering will be ever more intense and more dehumanizing.

Postscript - The Story of One Liberal Catholic Bioethicist

THE COMPLEXITY OF REMEMBERING

Scientists have recently made important contributions toward understanding changes that occur in the brain over time. In this chapter, I'll be exercising the neurons of my brain implicated in long term memory. My intention and commitment is to express my memories truthfully. I have to admit, however, that in light of recent scientific discoveries my recollections are something less than bits of objective data.

A memory is not something that is packed away in a safe place and then recovered at will. Memories are formed as a result of changes which take place in brain cells. At the molecular level a memory is continually influenced by other cellular events and other neuronal communications. There are billions of nerve cells (100 billion) and an intricate system of neuronal synapses. The synapses and intercommunications continually alter the inner cellular material involved in memory. Science has taken just the first steps in understanding this unimaginable complexity. We have much to look forward to. As science advances people will understand a little better why persons who experience the same life events in life often have very different memories of what happened and very different versions of the truth.

Our continuing experiences influence but do not totally change our memories. We still can remember certain childhood events which other family members corroborate. I have a memory of the events which I'll be trying to describe in this paper and many of these can be corroborated by friends who were with me at the time. Despite the cascade of changes involved in creating and then in storing a memory at the molecular level, a degree of accuracy in memory is possible that is nothing short of miraculous. The discoveries of science which are just data for some scientists, for me are the beginning of a sense of awe. I thank God that I am able to remember. I ask pardon for any imperfections in telling my story truthfully.

AN EARLY ROMAN EXPERIENCE

After I finished college at St. John's Seminary, the bishop of Little Rock, Albert Fletcher, sent me to Rome to study theology. Rome was supposed to provide a special kind of formation and to prepare future priests to become members of the hierarchy. Some of my classmates actually talked about how they were going to run their dioceses. Studying in Rome was their first step in a hierarchical career.

To me, such talk was a joke. I couldn't imagine becoming a priest in order to pursue a career in the Church hierarchy. To become a priest was to give oneself to God "not solemnly, not grudgingly, but to take one's life and throw it at God's feet and sing and sing; happy to bring you this small thing." Setting out to become a bishop seemed comparable to setting out to become a plant manager or school superintendent.

Being in Rome, however, and being around the Vatican, one could not miss the self-promoters and the career maneuverings. The Rector of the North American College frequently held get togethers for high level politicians and prominent hierarchy members. One evening there was a party attended by bishops and cardinals, diplomats and politicians. A friend from Iowa and I were given the job of opening the door, greeting the guests, and having them sign a book before they went upstairs for the celebration. After all the guests had arrived, and before returning to our rooms to study, my friend and I decided to sign the book ourselves.

Some days later Bishop O'Connor, the rector, called us to his room. He was furious. His face was red and he could hardly contain himself. "What right did you have to sign this book? What have you ever done for the Church? You have ruined my book! I ought to send you packing. Your bishop will hear about this." Wow! What a reaction! I don't remember feeling all that guilty or remorseful. What I do remember was thinking, "This guy is not living in the real world." "Who the hell does he think he is?" He takes himself far too seriously, I thought.

Over time I was more and more convinced that Vatican society not only has a human side but an aristocratic atmosphere of privilege and entitlement. Instead of making me interested in an ecclesiastical career, being in Rome did just the opposite. I thought to myself, I don't want to have anything to do with an ecclesiastical career. The last thing I would try to do would be to become a bishop.

That first experience with the upper class was the beginning of a perspective and an attitude which would endure throughout my life. It explains perhaps the liberal beliefs I developed and some of the trouble I got into with Church Officials. Any idea of looking at the Church hierarchy and the Vatican bureaucracy in deferential or reverential terms was gone. Instead of becoming a Church hierarch, I wanted to see changes in the Church hierarchy which would make it easier to reconnect with Protestant and Orthodox brothers and sisters.

Their problems with the Catholic Church I came to see had less to do with doctrine and more to do with politics. What split the Church in the 10th century and in the 16th century were the inexcusable behaviors of Church hierarchs and abuses of authority by Church bureaucracies. What will reunite the Church is reform of the Papacy, the Vatican, and the whole political system of the Church. This I came to believe early and continue to believe.

INTELLECTUAL LIFE AS A SEMINARY PROFESSOR

My early impressions about the political structures of the Church were strengthened by later studies at Protestant institutions and associations with Protestant scholars. Dr. James Gustafson was one of the strongest influences on me. My memory is not solid, but I think we met while I was studying at Union Theological Seminary in New York. So many Protestant and Jewish friends were such good persons that I quickly lost any sense of religious superiority. My interest was in closer relations with them and in the ecumenical movement. So whenever I could, I studied at non-Catholic institutions and I became comfortable in other religious cultures. Most often I was the only priest or the first priest to study in these settings.

After ordination, I was assigned to teach at St. John's Seminary. First I taught Latin, then modern languages. After some years the bishop sent me away to study philosophy at the University of Madrid.

Albert Fletcher, the bishop of Little Rock, was a good man, a decent and humble person who had none of the offensive characteristics of so many career-minded clerics at the Vatican. Even back in those early years, however, the bishop must have been somewhat concerned about me. I enrolled as a doctoral student in Madrid while the Council (Vatican II) was in session. Shortly, after the beginning of school, the bishop flew from Rome to Madrid to talk to my main professor and mentor, Jose Luis Arauguen. He wanted to make sure that I would get a straight Catholic formation. He went away reassured because Jose Luis Araguren was not only Spain's most prominent philosopher but a practicing Catholic and a deeply religious man who had written several important books on the Church in modern culture.

What the Bishop did not catch was that Jose Luis Aranguren, the Catholic philosopher, was also a most prominent advocate of a liberal Catholicism. He wanted a separation of Church and State in Spain. He wanted Protestants and Jews to have the same political status as Catholics. He wanted an end to the privileged connection of the Catholic Church with Spanish politics. Church establishment, he showed in his writings, had a deleterious effect both on the Church and on the Spanish nation. The doctoral dissertation which I wrote was on religious toleration and the demand for freedom of religion in a modern state. I brought to Spain the thinking of John Courtney Murray.

Both Jose Luis Aranguren and James M. Gustafson were major influences on my thinking and help explain both my liberal convictions and my activities as a priest. Jim Gustafson taught me that the Church had to be understood via biblical and theological categories but also with the aid of contemporary social sciences. Jose Luis Aranguren taught me that being a good Catholic meant taking responsibility for Church history and politics.

Anyone who lived during the 1960s had to recognize that there is a Catholic Church politics. The bishops at the Vatican Council were divided into political camps. The Vatican bureaucrats tried to force their views on local bishops. The Pope had his own core of political associates who made sure that his objectives and convictions were not undermined. The bishops divided into political groups and maneuvered with one another and with the Vatican politicians. The Vatican Council was more than a ritual get together of Church hierarchs. It was a hard fought political battleground. Only naïve Catholic fundamentalists see the Church without reference to its human dimensions, especially its strong political dimension.

THE ISSUE OF BIRTH CONTROL

One of the things going on during the Council was a Commission called by Pope John XXIII to look at the issue of birth control. That good and gentle Pope wanted a change in the Church's moral teaching on this subject because of enormous changes which had taken place in human population and because of the struggle Catholic couples were going through to raise a family responsibly. The Commission went on for years and I, like many others, followed as well as we could the Commission's deliberations. Here, too, politics and political struggles permeated the process.

Mr. and Mrs. Pat Crowley were two of only a very few Catholic lay members of the Commission. They stood for change after having surveyed members of the Christian Family movement which they headed. Catholic women expressed to them their terrible pain and frustration trying to raise their families without the aid of effective birth control. On the opposite side were some Vatican hierarchs who could not even imagine the possibility of change because, among other reasons, Anglicans and other Protestant Churches had already approved such changes. If the Church changed, it would appear that the Catholic Church was behind Anglicans and Protestants in moral teaching.

One especially narrow-minded Spanish theologian made the statement at one of the sessions: "What about the many people who are in hell because they violated Catholic moral teachings on birth control?" Mrs. Patty Crowley, who ordinarily was reserved and quiet at the meetings, as good laymen and laywomen were supposed to be, broke in after that remark with a devastating rebuttal. "Do you really think that God carries out your judgments about who goes to heaven and hell?"

After several years of deliberation, the Commission sent a final report to the new Pope (Paul VI) approving the use by married couples of effective methods to control procreation. (The pill had been discovered by a Catholic physician, Dr. John Rock and was seen as a godsend by Catholic and Non-Catholic couples alike). Only a very few Commission members (fewer than 5 of over 50 members) disagreed with the endorsement of change. They were Vatican bureaucrats joined with one U.S. Jesuit moralist, and a Cardinal from Poland (that Cardinal later became the Pope) who had never attended a meeting to listen to any of the arguments. This small group worked behind the scenes, after Commission members left Rome, to convince the Pope to reject the majority and to refuse to change. Pope Paul VI expressed their view in his infamous encyclical Humane Vitae.

It was some months before the publication of the encyclical that I got into trouble. I was fired from the seminary and suspended from the priesthood.

ONE FATAL ARTICLE

I have written hundreds of articles, but it turns out that one stands out as the cause of a major dislocation in my life. Most of my articles were published in academic journals and never caused the slightest ripple or response.

I was interested in the birth-control issue because of things I heard from women in the confessional and from couples who came to talk to me. In one instance, a few months after ordination, a young couple came to see me who already had a large family (6 or 7 children). Her periods were very irregular and every procreative control effort had failed. They were at wits end. I had learned well the traditional arguments against artificial birth control and explained the reasoning behind the traditional Church teaching. I convinced the husband, but not the wife. Then just a month or so later, someone told me that the couple had separated. I almost fainted. I was in shock. Did I contribute to the break-up of a marriage? I must have. That very day I started to re-examine the traditional teachings. Years later, the Arkansas Gazette published a product of that re-examination: the first public dissent from the official moral teaching about birth control. Then my life changed.

I had held the article in a drawer of my desk for more than a year. I knew that if I were to publish it, I'd be in trouble. (I thought they would punish me by sending me to an isolated parish in Slovaktown, Arkansas, which was located in the middle of a mosquito-infested rice field.) It was another meeting with another couple in trouble because of the absence of effective birth control that triggered my decision. After the couple left my room at the seminary, I opened the desk drawer, took the article, and went directly to the editorial department of the paper. Frankly, I thought that the article was too long and too academic to be of any interest. I thought the Gazette would reject it. Instead the editor of the editorial page saw its potential explosiveness, broke it down into op-ed lengths and published it in 5 separate pieces one day after another. Just a few days after the appearance of the last section, I received the letter from the bishop which fired me.

The priests at the seminary who were my age had all earned doctoral degrees at universities around the world at about the same time. We all caused the bishop to be nervous and uncomfortable. One close friend, Walter Clancy, had studied Canon Law and sociology and was involved in the Christian family movement. Dave Boileau had studied philosophy in Belgium and had written his thesis on racial segregation. Joe Biltz studied moral theology and was an advocate for social justice. Everett Ballman did contemporary scripture studies and came back with all sorts of new ideas about the Bible. My specialty was ethics, and the topic of my doctrinal dissertation was tolerance. We all became involved with one another's projects. We all wrote in the newspaper and were often consulted by radio, T.V. and newspaper reporters. We all took part in demonstrations and protests against the segregated Arkansas establishment.

After finishing my studies at Madrid, I had sneaked into the Soviet Union and upon my return, I wrote a book about that experience. Subsequently, I was condemned from the floor of the Arkansas Senate for being a Communist. The younger seminary faculty were all out on the edge as far as the bishop was concerned. But it was only after the articles on birth control that he cracked down. After firing me, he closed the seminary.

Ironically if birth control had been the only issue, the bishop might not have responded so strongly. But those earlier experiences with Vatican officials and Church political structures might help to explain a throw away last line at the very end of the birth control article. We had heard that there was a possibility that Paul VI might reverse the Papal Commission's recommendation. At the end of my article, I said something to the effect that a solemn Papal declaration condemning what the Papal Commission had recommended would have a hidden advantage. It would do away once and for all with the myth of Papal infallibility.

THE ISSUE OF INFALLIBILITY

The Infallibility idea had been muscled through Vatican I by Pius IX, the Pope who solemnly pronounced a series of condemnations in his Syllabus of Errors. Among the things that he condemned were democracy, freedom of speech and modern science. Pius IX in fact, condemned everything in sight, and tried to suggest that his reckless proclamations were supported by an infallible authority. He scandalized intelligent Catholics and left the rest of the world dumbfounded. He held the papal office for 32 years, one of the longest reigns in history (1846-1878). The longer he ruled, the worse things became.

Many bishops at the first Vatican Council opposed the idea of Papal infallibility, but they were politically outmaneuvered. They felt helpless in the face of Vatican political power at the Council and many simply left the Council rather than participate in what they saw as a rigged vote. They refused to vote because the process was unfair. One of the dissenters was the bishop of Little Rock, Arkansas. I was not the first rebel in that obscure, little diocese, where Catholics were less than one percent of the population.

Pius IX's Syllabus of Errors was part of a battle between the Papacy and the world which Pope, John Paul II, continued to wage. (The rule of John XXIII was a short relief from this mind-set.) Paul VI joined in this absurd confrontation between Papacy and the world when he condemned contraception even for married couples struggling to behave rationally with their families. The arguments had gone in the opposite direction in the Papal Commission. The Pope took his stand not based on rational arguments but on his own authority. If

Popes are infallible then the Church cannot change, he thought. Former Popes had said no to birth control (Casti Connubii 1930) and that settled the question.

If anything undermines the believability of the Catholic Church in a contemporary world, it is the idea that the Church cannot change. Everyone knows that change is inevitable both in the Church and everywhere else. Even moral teachings change because the realities which they address change. Once the Church endorsed slavery and condemned charging interest. The only thing more absurd than changelessness is the idea that Popes are infallible. Were the Borgia Popes infallible? Was Urban VII infallible when he condemned Galileo? Flat-footed, simple-minded understandings of Papal infallibility make the Church ridiculous. That's why I said what I said at the end of the article. This attitude is typical of liberal Catholics.

I regret many things which I have done in my life, but I don't regret trying to remove absurdities which undermine the Church's credibility. I remain a faithful member of the Catholic community. I had served on my parish board. I still go to mass as often as I can. Leaving the Church is not an option for me. My responsibility is to try to change what is wrong and not to stand by silently as the absurdities continue. I will continue to be involved in my Catholic community. The Church has done great good throughout the centuries and retains potential for continued good. The only way evils can be curtailed is to take up a fight for change.

By sheer coincidence, I had one more close association with Church hierarchy. Every year, North America College alumni gathered somewhere for a reunion. At one session I got up to entertain with a few friends. (I sing and dance.) On my way back to our table, an old priest stopped and asked my name and diocese. He said "We are going to do something for you." Thank you, I said, and started back to my table. Only then did I notice another old priest sitting next to him. It was Cardinal Spellman.

Later that year, I got a letter from the Cardinal inviting me to accompany him on a world tour. One of my companions on the trip was a Monsignor Cooke who later became the Cardinal Archbishop of New York himself. It was an interesting experience. We were dined and entertained by Presidents and Prime Ministers at every stop. In between stops, I had a chance to know many of the powerful lay Catholics from around New York and the U.S.A. There was satisfaction with my job performance, and I was told to keep in touch. But when I got to Little Rock, I decided not to do so. The idea of my becoming a priest was to give away my life, not to construct a career. I didn't think I would miss the culture and politics of hierarchical life.

LIFE OUTSIDE THE PRIESTLY CULTURE

After having been fired, my lawyer (Walter Clancy) and I initiated an appeal. We lost at the diocesan level, then again at the archdiocesan level and finally at the Vatican level. But then came an unexpected turn. I was in Rome at the time of the trial and one of the judges, a younger man, told me that he thought I had gotten a bad deal. He couldn't change the decision in favor of the bishop, but he offered me laicization. I took his offer and started another phase of my life but not before going through some very painful separation experiences.

Change is difficult for me. For example, I couldn't stand to see myself in a necktie. It made me sick to my stomach. And for the first time in my life, I was alone and completely broke.

I spent a few years at Yale with some help from a Jewish friend, and then took a job at Edinboro University of Pennsylvania. At the time, the university administration planned a broad expansion which was to include a medical school and extensive graduate programs. Over the years, that state institution became home for me, and I am still working from my office, The James F. Drane Bioethics Institute. Now my community is a secular university where many of my colleagues are fallen away Christians and Jews. I try to function as a priest for them. The Church, I've always heard, has a priest for every culture.

In the 1960's, while still at Yale, a friend named Dan Callahan and I went around the world. He was studying abortion legislation in different cultures and I was studying population control measures. All the time Dan talked about the possibility of leaving his job as editor of Commonweal and starting the first institute for the study of ethical issues in medicine. At that time, the issues were primarily ones of the beginning and end of life. On our return, he left Commonweal and founded the Hastings Center in Hastings, New York. The next year the Kennedy Institute was founded at Georgetown University and a discipline which came to be known as bioethics was born. Since then I have been working in this area. I still hold the Russell B. Roth professorship in Clinical Bioethics at Edinboro University. This professorship has a community-service dimension. Although retired officially, I help to organize and train ethics committees at hospitals and nursing homes in Northwestern Pennsylvania and throughout Latin America.

By coincidence or luck, I happened to have been standing by when bioethics as a discipline came to be. Then I happened to be part of an effort made by leaders of the Pan American Health Organization to start development of the discipline in Latin America and the Caribbean. I traveled throughout Latin America trying to identify resources which could help to develop bioethics there. I've spoken throughout the region at medical centers and with politicians interested in designing medical ethics legislation for their countries. A flourishing bioethics program is now established in Chile. Master's degree programs for the region are up and running. Institutes have been established for promoting bioethics in different countries. Dialogue with North American bioethicists is taking place, and a distinct Latin American bioethics is evolving under the influence of native-born scholars.

In 1990 while working as the resident bioethicist at the Pan American Health Organization, I attended a meeting of religious leaders from around the world on HIV. Among those in attendance were some African bishops who had studied in Rome about the same time I had. Of course we became friends immediately and they told me intimate details about the situation in their dioceses. A whole generation of their Catholic people had been devastated. Girls were afraid to marry because they never knew if the man was infected with HIV. There was not enough money even for testing.

Following this meeting I wrote an article on the use of condoms not for contraception but as a protection against the transmission of a deadly disease (i.e., until the young couple could be sure that each was free of disease). It appeared as a Cover Story in Commonweal. The next week I received a call from Donald Troutman, the bishop of Erie. He called to tell me that I had been removed from all committees at Catholic institutions. I asked what his objections were, but he did not say. It was enough, that I had touched the same old button which caused the first firing. (Ironically, later on Bishop Troutman got into trouble himself for trying to make needed changes in liturgical language. He received the standard authoritarian treatment himself. I wish him well, but I know that he will not ascend the career ladder. His career possibilities are gone.)

A LIBERAL AGENDA FOR CHANGE IN THE CHURCH

What we need as Catholics to advance the Church's credibility and effectiveness in the modern world is a more sophisticated epistemology. We have to get beyond the stupidity of always having to be right. If we do so, it will make other necessary changes easier. And it will help to put behind us the scandal of trying to cover up errors and acting as though they were somehow expressions of truth. There have been several attempts by the present Church administration to tell the truth because people demand this honesty.

After hundreds of years, there was finally some recognition of the problem with Galileo. (History shows that Galileo himself bears some responsibility for his censorship.) But the admission was so qualified that it came over as not all that honest. Then came recognition of failures for the Inquisition but again with nuances and qualifications so that it, too, came off primarily as exculpatory. Finally, the recent Papal document on Jews and the holocaust was perhaps the best example of the problem which Church authorities have in telling the truth.

Even when the Pope himself wants to tell the truth, the bureaucrats who write Church documents undermine it. It took them 10 years or so to produce the document about Jews and the holocaust. The final statement claimed that Catholics were guilty of participating in atrocities against Jews but not the Church, i.e. not the hierarchy. Of course this is a blatant untruth. What makes the Vatican create such scandal? The answer is clear. "It's infallibility, stupid."

Infallibility, in the mind of ordinary Catholics, means that the Church in the sense of the Pope does not err. Consequently every obvious error has to be covered up. Error cannot be admitted and therefore no sincere apology can be offered. Every movement to change this crazy system has had little success. As long as Popes like John Paul II and Bendeict XVI are in charge, success will be limited at best. They will go to any length to bolster the Papal infallibility idea.

The continuing emphasis on infallibility pleases the ultra-conservatives but to most people in most cultures, it is absurd. Was St. Peter infallible when he made mistakes and St. Paul confronted him at Antioch? During the long history of the Church, the list of errors in solemn Papal pronouncements is a long one. Right wing theologians can find subtleties and distinctions to excuse the errors much like the contemporary Vatican theologians excuse the Church of any responsibility for the Nazi holocaust.

Under a less authoritarian leader, the Church hopefully can take some steps to move away from the present political system, permeated with the practice of covering up the truth. Many Protestant and Orthodox believers would feel comfortable about approaching closer union with the Catholic Church if more truth and honesty and less righteousness and cover-up were reflected in the Church's public behaviors. There can be a papacy which maintains authority and which symbolizes Church unity, as long as this office abandons authoritarian patterns and removes the absurdity of having a monopoly on truth. Instead of covering-up and denying truth, the papacy we need is one that stands for truth: not a stingy, minimal, qualified truth, but a generous, full and total truth. But such an ideal is very hard to attain. Maybe Catholic liberalism or at least this Catholic liberal is unrealistic.

TELLING THE TRUTH

Telling the truth about anything as complicated as history is a major challenge. Truth has to be framed, which means that it has to be formatted in some way. Many of the Church's critics suffer from a pathological anti-Catholicism and it comes out in their framing historical events from the most negative slant possible. Anti-Catholic prejudice is still alive and well in many of the main-line Protestant Churches. Even liberal criticisms of the Church's flawed attempts to tell the truth can themselves be seriously flawed. Many judge past events by present values, and this is both illogical and unfair. Values, too, pass through eons of time and epochs of social change. Lacking this basic understanding, people cannot morally evaluate either present reality or historical contexts. Serious critics of the Church's moral teachings and official Vatican declarations do not want Church history judged by present values. Rather they want an honest admission of past mistakes and they want the mistakes judged both in past and present contexts.

Truth can be told, even difficult truth about complicated historical periods. But to do so we can begin neither with deep-seated prejudices nor with isolation in our present little world. We have to begin by sharing the social space of the events that are being evaluated and judged. Knowledge is historical. Truth is historical. And values, like all other reality, go through an evolutionary process. These are basic liberal Catholic suppositions. They are the very antithesis of tribal attitudes and dispositions that are so widespread today.

Tribalism has re-emerged in the post cold-war period and the Church is not immune from its influence. The tribal group or Church or ethnic club becomes a place for attaining a facile security and status. Both are achieved simply by being a member of the tribe and wearing the tribal symbols. Tribal identity is wrapped in self-righteousness, and deep-seated rivalry or prejudice against other groups. The tribal mentality has nothing to do with the mind-set of Jesus and nothing to do with his directives for his followers. Jesus opposed the Jewish leaders and their tribal interpretations of God's revealed truth.

And yet, tribalism has entered the Church culture. Many Church leaders, and not just Catholic ones, act like the Jewish authorities who thought they knew everything about revealed truth. But no truth about God can adequately be expressed by human beings, in human terms. We use images and stories and concepts and doctrinal statements but all these fall short of the whole truth. Why do today's religious leaders think they have the unchangeable truth? Why do they adopt the behavior of the leaders whom Jesus refused to follow? Why then do people fight and kill each other over imperfect expressions of religious truth? Because people in the past and people today easily slip into tribalism. They think their little belief statements capture the whole reality and all others are wrong. And they tend to be moral tribalists as well. They believe that only their group is right. All the others are judged not by what is best about them, but by what is worst about them. Tribal types have to be right. Just listen to spokesmen for the simpleminded militia groups and look at the loss of innocent life they excuse as collateral damage.

Islam and Judaism for example, insist that God is One. Christians agree but also hold that God is Trinitarian and relational. But all such talk is metaphoric. It's never the whole truth. We cannot restrict God to the limits of human language and especially not to Vatican or Islamic or Jewish proclamations. What is true of dogmatic proclamations is even more true of moral teachings. The idea that a papacy could define the morality of a concrete act like birth control, about which scripture says nothing, and do so based on a defective Natural Law vision joined to a claim of infallibility is so absurd that it demands to be publicly repudiated.

Especially in the area of history and religion and morality, telling the truth is never a matter of syllogisms, or pure deduction. Even subtle expressions of truth are never the whole truth. Truth is always complicated. Catholics have to insist on the truth. We have to work to achieve the truth. We have to demand an end to lying. But we have to hold the Church and ourselves to a mature rather than to a simplistic epistemology.

Liberal Catholics want the Church to tell the truth. They want a truly Catholic Church, and not a tribal Catholicism. But, truth telling is a challenge, especially in religion. The self-righteous and the simple minded do not confront such challenges. They know exactly what is true and their convictions are like physical laws: they never change. Their experience confirms their righteousness because they line others up according to tribal standards: us versus them. Tribal types are either the good guys or the victims. One thing they cannot abide or even consider is another point of view or a different conviction about the same issue.

As for the Church, one simple little change could go a long way toward avoiding tribalism and reducing the stream of scandal which has flowed from the ultra-conservative papacies of Pius IX and John Paul II. Both Popes have caused the Church harm by being in that Papal chair too long. We need a 10 or 15-year term restriction on the Popes. Pius IX was Pope for far too long and so was John Paul II. Both men became celebrities and icons for fundamentalist admirers. This is always dangerous. We need term limits. That would save us from being subject to some of the worst Papal policies. It would also make it more likely that we would get to know the truth and be saved from the scandalous cover-ups. Finally we would be saved from practices of meanness toward faithful liberal members of the Church who disagree with the authoritarian behaviors of celebrity Popes.

My liberal Catholic perspective permeates what I have written about contemporary bioethics issues. It shows most clearly in the treatment of certain issues, but it is always there and always an influence. Although I feel strongly about the issues, I am always aware of the fact that mine is only one perspective and that I am subject to the fallibility which accompanies ethical arguments about complex issues. I may argue strongly, but I know that even my strongly held convictions are fallible. We do not need final infallible moral teachings, but rather a Church environment where continued open dialogue is possible. If this book is used in bioethics classes, it will be one perspective among others which students can discuss and then decide about. If the book is used by medical professionals and health-care administrators, perhaps the arguments could lead to the establishment of more liberal institutional policies. In either case, I would be pleased. It would be satisfying to know that what I worked out in my own mind over the years became one voice in the public dialogue, at least for a few years.

INDEX

A

B

I

J

K

L

U

V

W

NOTES

DEFINITIONS

What is More Humane Medicine?

[1] cf. Kevin Wilder S.J. "Patient no more," America, July 16, 2001
[2] The importance of traditional medical ethics is emphasized by Edmund Pellegrino, cf. Edmund Pellegrino, The Philosophy of Medicine Reborn, (University of Notre Dame Press, 2008) and Physician and Philosopher: The Philosophical Foundations of Medicine: Essays by Dr. Edmund Pellegrino, (Garden Jennings Publ., 2001).
[3] For a more developed treatment of the place of virtue and character in bioethics, cf. Drane, James, Becoming a Good Doctor, (Kansas City, MO. Sheed and Ward, 1995).

What is a Liberal Catholic?

[4] A more modern understanding of what Liberal Catholic means is provided by a study entitled Catholicism on Campus. The study was conducted by CARA (Center for Applied Research in the Apostolate). It reported that students at Catholic Colleges and Universities adhere most strongly to liberal teachings and perspectives. These more liberal young people are the ones most alienated by official Church stands on morality. The study was released January 30, 2010.
[5] cf. J.C. Murray, We Hold These Truths: Catholic Reflections on the American Proposition, (Rowman & Littlefield Pub, 2005). Religious Liberty: Catholic Struggles with Pluralism, (Library of Theological Ethics), (London: Westminster John Knox Press, 1993). Problem of God: Yesterday and Today, (New Haven, CT: Yale University Press, 1964). Bridging the Sacred and the Secular: Selected Writings of John Courtney Murray (Georgetown University Pub, 1994).
[6] cf. J. Noonan The Luster of Our Country: The American Experience of Religious Freedom, (Berkeley, CA: University of California Press, 1998). Contraception: A History of Its Treatment by the Catholic Theologians and Canonists, (Cambridge, MA: Harvard Univ. Press, 1986). The Morality of Abortion: Legal and Historical Perspectives, (Cambridge, MA: Harvard Univ. Press, 1970). Church That Can and Cannot Change: The Development of Catholic Moral Teaching, (University of Notre Dame Press, 2005). Religious Freedom: History, Cases, and Other Materials on the Interaction of Religion and Government, (Foundation Press, 2001).
[7] In September 2000, The Atlantic Monthly ran an article entitled, "Fourteen" by Stephen Zanichkowsky. His recounting of his family history would make any sensitive person cry and then move to insistence on birth control for preserving family health and sanity.
[8] G. Wills, Papal Sins, the Structures of Deceit, (New York: Doubleday Press, 2000). Subsequently he published, Why I Am A Catholic, (Houghton Mifflin, N.Y., 2002). Under God: Religion and American Politics, (Simon & Schuster, 2007). Head and Heart: A History of Christianity in America, (Penguin Group, 2008).
[9] J. Carroll, Constantine's Sword: The Church and The Jews: A History, (New York: Houghton Mifflin, 2001). Toward a New Catholic Church: The Promise of Reform,

(Houghton Mifflin Harcourt, 2002). Practicing Catholic, (Houghton Mifflin Harcourt, 2009).

[10] cf. Spiritual Questions for the Twenty-First Century, ed. Mary Hembrow Snyder, (Maryknoll, NY: Orbis Books, 2001). Gift of Years: Growing Older Gracefully, (Bluebridge, 2008). Joan Chittister: In My Own Words, (Liguori Publishing, 2008). Rule of Benedict: A Spirituality for the 21st Century, (Crossroads Publishing Company, 2010). Uncommon Gratitude, (Liturgical Press, 2010); This book was published with rowan Williams, the Anglican Archbishop of Canterbury.

[11] For a more conservative perspective, cf. Edward J. Furton, Catholic Healthcare Ethics: A Manual For Practitioners, (St. Augustine Press, 2009).

What is Bioethics? A History

[12] V. Potter, Bioethics, Bridge To The Future, (Englewood Cliffs, NJ: Prentice Hall, 1971).

[13] Fritz Jahr, Bio-Ethik. Eine Umschau über die ethischen Beziehungen des Menschen zu Tier und Pflanze. Kosmos. Handweiser für Naturfreunde, 1927, 24: 2-3. - For different views of the beginning of Bioethics, cf. Erin P. Klein, Jennifer K. Walton (editors), The Story of Bioethics, (Georgetown University Press, 2003). George Annas, American Bioethics: Crossing Human Rights and Health Law Boundaries, (Oxford University Press, 2009). Nancy Jecker, Albert Jonsen (editors), Bioethics: An Introduction to History, Methods and Practices, (Jones and Bartlett Learning, 2007). Bernard Gert, Charles Curran, Bioethics: A Systematic Approach, Oxford University Press, 2006). Hans-Martin Sass, Fritz Jahr's Early 1927 Bioethical Imperative. Kennedy Institute of Ethics Journal 2007, 17(4):279-295

[14] Nuremberg Code. Cited by R. Levine, Ethics and Regulation of Clinical Research, Second Edition, (New Haven, CT: Yale University Press, 1988).

[15] H. Beecher, "Ethics and Clinical Research," The New England Journal of Medicine 274, (1966), 1354-1360.

[16] Later, Jay Katz published important books and articles on ethical aspects of the doctor/patient relationship.

[17] New technologies developed after WWII launched contemporary American medicine. Continuing streams of new technologies now threaten the American healthcare system. Cf. Daniel Callahan, Taming the Beloved Beast: How Medical Technology Costs are Destroying Our health Care System, (Princeton University Press, 2009).

[18] The previous Council members' articles were also published. Cf. Leon Kass, Being Human: Core Readings in the Humanities, (WW Norton & Co Inc., 2004).

[19] Articles of the Council members were published. Cf. Edmund Pellegrino, Human Dignity and Bioethics, (Government Publication, 2008).

[20] Idea of Care: A History and Analysis, (2010).

[21] Contemporary Issues in Bioethics, (Cengage Learning, 2007). Editor, Source Book in Bioethics, (Georgetown University Press, 2000).

[22] Standing of Principles: Collected Essays, (Oxford University Press, 2010). Principles of Biomedical Ethics, (Oxford University Press, 2008).

[23] Editor, Belmont Revisited: Ethical Principles for Research With Human Subjects, (Georgetown University Press, 2005). Principles of Biomedical Ethics, (Oxford University Press, 2008).
[24] Clinical Ethics: A Practical Approach to Ethical Decisions in Clinical Medicine, (McGraw-Hill, 2010). Short History of Medical Ethics, (Oxford University Press, 2008).
[25] Clinical Ethics: A Practical Approach to Ethical Decisions in Clinical Medicine, (McGraw-Hill, 2008). *La importancia del aspect practico y clinic de la etica en la formacion de los profesionales de al salud. La experiencia del McLean Center for Clinical Medical Ethic*, Educar en Bioetica al profesional de Ciencias de la Salud. Una perspectiva internacional, (Imprimex, 2010).
[26] Global Bioethics: The Collapse of consensus, (M&M Scrivener Press, 2006). Editor, Bioethics and Moral Content: National Traditions of Health Care Morality, (Springer Verlag, 2003).
[27] Taming the Beloved Beast: How Medical Technology Costs Are Destroying Our Health Care System, (Princeton University Press, 2009). Medicine and the Market, (Johns Hopkins University Press, 2006). What Price Better Health?: Hazards of the Research Imperative, (University of California Press, 2005).
[28] *Corn v. French*, 289 P.2d 173 (Nev. 1955); *Wall v. Brim*, 138 F.2d 478 (5th Cir. 1943); *Waynick v. Reardon*, 72 S.E.2d 4 (N.C. 1952); *Nolan v. Kechijian*, 64 A.2d 866 (R.I. 1949)
[29] Salgo v. Leland Stanford Jr. Univ. Bd. of Trustees, 317 P.2d 170 (Cal. Ct. App. 1957)
[30] *Natanson v. Kline*, 350 P.2d 1093 (Kan. 1960)
[31] J. Drane, Clinical Bioethics, (Kansas City, MO: Sheed and Ward, 1994). A later book continues Bioethics instruction for ethics committee members, James F. Drane, Caring to the End, (Lake Area Health Education Center (Erie, PA), 1997).
[32] S. Toulmin, "How Medicine Saved the Life of Ethics," Perspectives in Biology and Medicine 25 (Summer, 1973) 4: 736-750.

THEORY

Natural Law, History and Politics

[33] Cf. Alfonso Gomez-Labo, Morality and Human Goods: An Introduction to Natural Law Ethics, (Georgetown University Press, 2002).
[34] For St. Thomas, the hierarchical view of the natural order was self-evident. A more contemporary, scientific model, however, gives a very different, rational view of the natural order, and a different content of natural law principles. Beginning in the late Medieval and early Renaissance, the present more importance is given to human will and human intervention into the natural order. Cf. David J. Collins S.J. Ph.D., God and Nature in Western Thought, (Now You Know Media, 2009).
[35] cf. R. Niebuhr, Christ and Culture, (New York: Harper and Row, 1951), p. 145. Niebuhr makes this same point in his book, The Responsible Self, An Essay in Christian Moral Philosophy, (NY, NY: Harper and Row, 1963).

[36] The role of Eleanor Roosevelt in the Universal Declaration of Human Rights is the subject of a recent book. cf. M. Glendon, A World Made New, (NY, NY: Random House, 2001).
[37] This traditional view is challenged by some contemporary thinkers. Cf. Andrew Lustig, Baruck Brody (editors), Altering Nature: Concepts of Nature and the Natural in Biotechnology Debate, (Springer Verlag, 2008).
[38] G. Vico, The New Science, trans. T. Bergin and M. Fisch. (NY, NY: Anchor Books, 1961). par. 331 et al.
[39] E. Cassirer, An Essay on Man, (New Haven, CT: Yale University, 1951). Cassirer examines the major forms of human enterprise, i.e., history, art, myth, science, etc, showing these cultural expressions to be founded in the unique, symbolic nature of man.
[40] J. C. Murray, We Hold These Truths, (Barnhart, MO: Theological Book Service, 1986), p. 33. The book consists of a series of essays by the late Fr. Murray on issues facing a pluralistic society, the solutions to which Murray works out by using the methodology of Natural Law. The last essay, "The Doctrine Lives" is an important contemporary statement of Natural Law Theory.
[41] The historical roots of human being as free and self-creative are biblical: It was Abraham who defined himself by his free decisions and who initiated the inclusion of freedom into the definition of human being. It took many centuries before this freedom was politically recognized in documents of the French and American revolutions and included in a Catholic Natural Law vision.
[42] The worst/best examples of this perspective we find in the ultra Orthodox Catholic societies working to return to the Church culture of pre-Vatican II days, and in the continuing inquisitorial practices of some Vatican offices.
[43] cf. St. Thomas Acquinas, Nicomachean Ethics, Book 1, 3. "What pertains to moral science is known mostly through experience." St. Thomas makes the same point about the changeableness and developmental character of Natural Law in his treatment of the subject in Summa Theologica, First Part of the Second Part, Q 94, Art. 2, 4, 5. For other perspectives on St. Thomas, cf. John Goyette (editor), St. Thomas Aquinas and Natural Law Tradition: Contemporary Perspectives, (Catholic University Press, 2004).
[44] H. Theilicke, Theological Ethics, (Philadelphia, PA: Faters Press, 1966), p. 429.
[45] cf. J. L. Aranguren, Etica y Politica, (Madrid, Spain:Ediciones Guadarrama, 1963).
[46] G. Hegel, The Phenomenology of Mind, (NY, NY: Harper Torch Book Edition, 1967).
[47] For the scholastics, all of whom preoccupied themselves with the question of law generally and Natural Law in particular, the precepts of *Jus Gentium* were considered part of the Natural Law and valid for the ordering of the community of peoples. *Jus Gentium* was the quasi-positive law of the international community. Its fundamental axiom was pacta sunt servanda and it covered areas such as war, truces, trades, treaties, envoys, etc.
[48] For an interesting account of the situation of both Arabs and Jews in medieval society, cf. A. Castro, The Structure of Spanish History, trans. E. L. King, (Princeton, NJ: Princeton University Press, 1954). The original Spanish title was Espana en su Historia (Christianos, Moros y Judios).

[49] A famous debate took place in Spain (Vallodolid) between Bartolome de Las Casas, a Dominican friar known as the Apostle to the Indians and Juan Gines de Sepulveda. The year was 1550. Las Casas had written an account of the cruelties visited upon the Indians by Spanish colonizers. His work caused a controversy which led to the debate with Sepulveda, a famous classical scholar. It was the latter's contention that the Spaniards had a right to subjugate the savages because they were a lower order of nature. Las Casas refuted these arguments with others based on the Scriptures and on rational arguments taken from St. Thomas and others. Las Casas saw in the Indians signs of humanness which demanded decent treatment, even though they were uncivilized and unchristianized. Oddly enough, among the many concrete suggestions he made, one was to bring Africans as slaves to take over the tasks which the Indians were being required to do. cf. Tratados de Fray Bartolome de Las Casas, (Mexico-Buenos Aires: Fondo Cultura Economica, 1965).

[50] Grotius is often hailed as the Father of Natural Law. At a time when the religious solidarity of Europe was destroyed, he tried to substitute an intellectual solidarity. He tried to introduce the rule of law even in wartime (the Thirty Years War, 1618-1648) and after the religious foundations of civil peace had crumbled. cf. De Juri Belli et Pacis, (1925) which was reproduced photographically and translated by Francis W. Kelsey and others for The Classics of International Law, (London: Oxford Univ. Press, 1917).

[51] The issue of the togetherness of different people for contemporary Americans is addressed by M. Marty, The One and The Many: America's Struggle for the Common Good, (Cambridge MA: Harvard University Press, 1997). The different peoples in today's America are many. The challenge to togetherness and indeed to the survival of American society is tribalism. Race, language, gender, ethnicity, economics, nationality, culture and religion tribally organized and lived are all potential enemies of civic oneness and social togetherness. Marty addressed these challenges and the steps which Americans who make up these groups need to take in order to stay one nation; free, communicating, and sharing with one another.

[52] cf. W. Bosch, Judgment on Nuremberg, (Chapel Hill, NC: University of North Carolina Press, 1970).

[53] The primary thrust of Stoicism was toward the creation of individual self-sufficiency and virtue. The metaphysics of Stoicism, however, provided the foundation for a world-state idea and a corresponding International Law which was a Natural Law. They believed in the oneness of all nature and the identification of God and reason and nature. All men then were the sons of God and belonged to a world community which transcended political divisions. Right reason, teaching humans what is right and wrong, was the constitution of the world society. By the law or by reason all men were equal, slaves and free, Greeks and barbarians. Although social reform was a secondary consideration of the Stoics, their idea of the Natural Law was potentially a ground for social reform and improvement. Chrysippus insisted that slaves should be treated as a "laborer hired for life" which did represent an advance over Aristotle's idea of the slave as a living tool.

[54] Pius IX's Syllabus of Errors is an example of the reactionary function. His long list of errors were judged to be erroneous based not on scripture, but on human reasons' conclusions about morality in the social order. His reactionary view of Natural Law

morality was considerably influenced by the fact that Enlightenment governments stripped him of his Papal states and confined him to semi-imprisonment in the Vatican.
[55] Pope Paul VI attracted extensive international attention to the use of Natural Law Theory with his much-criticized encyclical Humane Vitae.

Natural Law and Universal Standards

[56] The new international legal guideline is entitled the Princeton Principles on Universal Jurisdiction. The principles are 14 in number and cover all aspects of international trials from definitions of crimes to disputes between governments over the accused.
[57] Daniel Maguire, a respected and widely published Catholic moral theologian has been targeted by another Catholic, Patrick Reilly, for his defense of woman's rights and his opposition to Vatican politics at international meetings. Instead of respecting the need for open discussion of complex and controversial issues (a long Catholic tradition) and the reality of diversity within the international Catholic community (an obvious fact), Mr. Reilly wants to clear the Church of diversity and to punish Catholics who disagree with his conservative views. Cf. Ethics: A Complete Method for Moral Choice, (Augsburg Fortress Pub, 2009).
[58] President George W. Bush reflects this idea in his political decisions to withdraw from all kinds of international treaties because they do not promote his view of national interests.
[59] This difficult proposition however is being advanced by Hans Kung, a liberal Catholic Theologian. He organized a Parliament of The World's Religions in Chicago in 1993 and has published a number of books on this topic: Global Responsibility: In Search of a New World Ethic, (London: SCM Press, NY, NY: Continuum, 1991), A Global Ethic: The Declaration of the Parliament of the World's Religions, (London: SCM Press, NY, NY: Continuum, 1993), and Yes to a Global Ethic, (NY, NY: Continuum, 1996). His is an ethic of four basic principles derived from world religions.
[60] The Nuremberg Code, (Washington D.C.: U.S. Government Printing Office, 1947).
[61] Later on, the same ethical standards used to judge Nazi doctors were applied to condemn similar technological abuses by U.S. physicians. News of grossly unethical behavior by some Nazi physicians during World War II was followed in the U.S. by a series of revelations of similar ethical failures involving experimentation on vulnerable patients in American medicine (Willow Brook School, Jewish Hospital in New York, and the Tuskegee Syphilis Study). In 1966 Henry K. Beecher, a Harvard physician, published an article in the New England Journal of Medicine in which he exposed patterns of unethical conduct in medical research.
[62] The U.N. Universal Declaration of Human Rights (1948) joined to the International Covenant on Civil and Political Rights (1966); The Declaration of Helsinki (1964) promulgated by the World Medical Association and revised in 1975 (Tokyo), 1983 (Venice), 1989 (Hong Kong); The International Ethical Guidelines for Biomedical Research Involving Human Subjects (1982)(1993).
[63] The Nuremberg Code (1947) set the standard of free and informed consent by competent subjects/patients. Over the years the elements of information, free consent,

and competency, essential for ethical research, were applied as well to patient involvement in treatments procedures. Informed consent constitutes the foundation of modern medical ethics, both research and clinical, world-wide.

[64] At the time of its final passage the most frequent violator and main source of opposition was the Soviet Union. Violators today continue to be in totalitarian and fundamentalist religious states. A recent report from China documents the use of psychiatry by the government to stifle political dissent. Dissenters are given electroshock and powerful medicines for what officially is referred to as paranoid psychosis. International medical associations and human rights groups have begun an international campaign to gain access to psychiatric hospitals and to the patients confined there for dissenting.

[65] Virtually every developed nation today has some type of National Ethics Committee which sets policies and proposes statutory regulations to legislatures.

[66] The following are examples of ethical codes and declarations which claim international status. They come from different sources: from Amnesty International, *The Declaration of Stockholm on the Prevention of Torture*; from the U.N., Principles of Medical Ethics; from WHO, *Declaration of Geneva, International Code of Medical Ethics*; from the World Psychiatric Association, *Declaration of Hawaii*; from the International Council of Nurses, *Statement on Nurses and Torture*.

[67] Convention for The Protection of Human Rights and Dignity of the Human Being with Regard to the Application of Biology and Medicine: Convention on Human Rights and Biomedicine. The Council of Europe. Strasbourg, Nov. 1996.

[68] The term bioethics was not used in the document because in the minds of some, it was identified with the opinions of Peter Singer, an Australian who endorses active euthanasia for declining elderly patients and defective infants.

[69] For a review of the involvement of medical professionals in violations of international ethical standards and human rights in China today: cf. "Psychiatric Abuse by China," New York Times, 18 February 2001.

Natural Law and Sexuality

[70] Soon after penicillin and even more powerful antibiotics, a U.S. Surgeon General announced that the book on infectious disease had been closed. Today, infectious diseases are even more threatening than they were before, because of antibiotics.

[71] Thirty thousand Americans kill themselves each year and a half-million attempt suicide seriously enough to require emergency room treatment. The most common reason is anger and aggression turned toward oneself. cf. K. Redfield Jamison, Night Falls Fast, (NY, NY: Alfred Knoff, 1999), p. 24.

[72] Menninger School of Psychiatry, one of the leading psychiatric institutions in the world, has a training program that targets the risk of suicide in psychiatric patients. Eight to ninety percent of psychiatric patients attempt suicide. For information about this program see Menninger Perspective, November 1, 2008.

[73] For example, Paul VI Humanae Vitae, Karol Wojtkla, Love and Responsibility, John Paul II, Theology of the Body: Love in the Divine Plan.

[74] St. Thomas Aquinas, Commentary on the Nicomachaen Ethics, trans. C.I. Litzinger, (Chicago, IL: Reginary, 1964).

[75] Rock, John, The Time Has Come: A Catholic Doctor's Proposals to End the Battle over Birth Control, Alfred A. Knopf: NY, 1963.

[76] Jonsen, Albert R. and Hellegers, Andre E. "Conceptual Foundations for and Ethics of Medical Care," in Ethics of Health Care: Papers of the Conference on Health Care and Changing Values, November 27- 29, 1973, pp. 3-20. Edited by Laurence R. Tancredi. Washington, D.C.: National Academy of Sciences. Albert Jonsen, one of the founders of Bioethics, continues his work in the discipline. In June 2010, at an International Congress on Bioethics organized by The James F. Drane Bioethics Institute at Edinboro University of Pennsylvania, Dr. Jonsen spoke on the topic, *Neonatology: Where Clinical Bioethics is Born*. The 7th edition of his book, Clinical Ethics: A Practical Approach to Ethical Decisions in Clinical Medicine came out in 2010. A Short History of Medical Ethics was published by Oxford University Press in 2008.

LIFE

HIV/AIDS and Condoms

[77] The lesser evil point, however, was recognized by Monsignor Sandeau, the Vatican moral theologian, in an article on AIDS in Observatore Romano. (April, 2000). Between 2000 and present (2010), several African Bishops have refused to condemn the use of condoms in marriage when on partner is already infected and the other is in danger of becoming infected. No changes from Vatican Authorities however have been made public.

[78] cf. J. Fuller and J. Keenan, "Tolerant Signals," America (September 23, 2000).

Papal Authority and Birth Control

[79] Today these figures are more like 90% and 95%.

[80] Roma locuta est, causa finita est, is the classical Latin formula for this idea.

[81] This statement certainly applies to Pope John Paul II's teaching about sexuality in Love and Responsibility.

[82] "The present Pope" refers to Paul VI.

[83] Some time after the publication of this article in the *Arkansas Gazette*, Paul VI published the Encyclical, Humanae Vitae.

Modifying Abortion Policies: The Role of Metaphor in Medicine Law

[84] The move towards replacing the hippocratic metaphor with today's scientific/ mechanic metaphor began with the physician Paracelsus in the 16th century. Cf. David Collins, God and Nature in Western Thought, (Now You Know Media, 2010).

[85] c.f. W. Austin, "The Fiduciary Principle," (1949) 37 Cal.L.Rev. 539-40.

[86] St. Luke's Gospel 16:10-13.

[87] American Medical Association, Report of the Council on Ethical and Judicial Affairs, Report A (I-86): Conflicts of Interest 11, (1986).

[88] ibid.

[89] ibid.
[90] American College of Physicians, Ad Hoc Comm.. on Medical Ethics, "American College of Physicians Ethics Manual, Part 1", Annals of Internal Medicine 101 (July, 1984): p. 129-137 .
[91] E. Pellegrino, et al, Ethics, Trust and the Professions, (Washington D.C.: Georgetown University Press, 1991); E. Pellegrino, For the Patient's Good, (London: Oxford University Press, 1987); E. Pellegrino, Health Care Professionals Friend and Healer, (Georgetown University Press, 2000).
[92] In the latest development in the U.S., a large pharmaceutical manufacturer purchased and then took over the management of 11 cancer centers. Independent physicians, who did not follow directions from the management about medical practice, found themselves out of work.
[93] A. Capron, "Anencephalic Donors: Separate the Dead from the Dying," Hastings Center Report 17 (February 1987), 1:5-9.
[94] K. O'Rourke, "Kidney Transplantation from Anencephalic Donors (Letter)," The New England Journal of Medicine 317 (October 8, 1987), 15:960-961. Robert Veatch, Transplantation Ethics, (Georgetown University press, 2002).
[95] Cranford, R. and Roberts J., loc. cit.
[96] N. Fost, "Organs from Anencephalic Infants: An Idea Whose Time has Not Yet Come," Hastings Center Report 18 (Oct/Nov, 1988), 5:5-10. Thomas Beauchamp, Contemporary Issues in Bioethics, (Rengage Learning, 2007).
[97] Shewmon, D. et al., loc. cit.
[98] Task Force for the Determination of Brain Death in Children, "Guidelines for the Determination of Brain Death in Children" Annals of Neurology 2 (June 1987), 6:616-617. (Same guidelines also found in Archives of Neurology, 44, (June 1987), 6:587-588; Pediatrics 80 (Aug. 1987), 2:298-300; and Neurology 37 (June 1987), 6:1077-078).
[99] M. Harrison and G. Meilaender, "The Anencephalic Newborn as Organ Donor (Case study and commentaries)," Hastings Center Report 16 (April, 1986), 2:21-23. James F. Childress, Organ Donation: Opportunities for Action, (National Academy Press, 2006).

DEATH

Aging and Dying

[100] P. Peterson, Gray Dawn: How the Coming Age Wave will Transform America and the World, (New York: Random House, 2000).
[101] A different analogy from Gray Dawn was used by Laura L. Carstensen in an op. Ed piece in the N.Y. Times, Jan. 2, 2001. She wants to redesign old age so that aging can be healthy and the elderly can make positive contributions to society.
[102] J. Aranguren, La Vejez Como Autorrealizacion Personal y Social, (Madrid, Spain: Instituto Nacional de Servicios Socials, 1992).
[103] In Hippocratic medical theory, the 7-year cycles were correlated with the four humors (black bile, yellow bile, phlegm and blood) and the 4 elements (fire, earth, air, water).

---

[104] For a recent response to this problem see, James F. Drane, Finding Relief From Suffering and Depression, (St. Paul's Press, 2010).
[105] cf. D. Callahan, Setting Limits, (New York: Simon and Schuster, 1987). Troubled Dream of Life, (New York: Simon and Schuster, 1993). False Hopes, (New York: Simon and Schuster, 1998). Taming the Beloved Beast: How Medical Technology Costs Are Destroying Our Health Care System, (Princeton University Press, 2009). Medicine and the Market: Equity V. Choice, (Johns Hopkins University Press, 2006).
[106] cf. J. Drane, Clinical Bioethics, (Kansas City, MO: Sheed and Ward, 1994), p. 156. Caring to the End, (LAHEC, 1997). Clinical Bioethics, (Sheed & Ward, 1994). Becoming a Good Doctor, (Sheed & Ward, 1995).

Euthanasia and Physician-Assisted Suicide

[107] J. Pat McCarrick, Active Euthanasia and Assisted Suicide, (Kennedy Institute of Ethics, 1992).
[108] G. Gruman, "Death and Dying: Euthanasia and Sustaining Life: Historical Perspectives," cited in Encyclopedia of Bioethics (New York: Free Press, 1978), p. 261-268; B. Fry, "Active Euthanasia: A Historical Survey of its Conceptual Origins and Introduction into Medical Thought," Bulletin of the History of Medicine 52 (Winter, 1978): 492-502.
[109] Phillippe Aries' books published in the 1980's are examples of studies of attitudes regarding death and dying.
[110] Don V. Bailey, Challenge of Euthanasia: An Annotated Bibliography, (Rowman & Littlefield, 1990).
[111] Jonathan d. Moreno (editor), Arguing Euthanasia: The Controversy Over Mercy Killing, Assisted Suicide and the Right to Die, (Simon & Schuster, 1995).
[112] For a concise and intelligent history of euthanasia, see Garcia, Diego, "Historia de la Euthanasia" La Eutanasia y El Arte de Morir, ed. Jacier Gafo S.J (Madrid: Universidad de Comillas, 1990), p. 13-32. Diego Gracia, Voluntad de comprension, (Triacastela, 2010). Lan Dowbeggin, concise History of Euthanasia: Death, god and Medicine, (Rowman and Littlefield Pub Inc, 2007).
[113] Carrick, Paul, Medical Ethics in Antiquity, (Dordrecht: D. Reidel, 1985), p. 136.
[114] The Republic, (Oxford, England: Oxford University Press, 1973), p. 97.
[115] Nicomachaen Ethics, trans. W.D. Ross, Basic Works of Aristotle (New York: Random House, 1941) p. 977.
[116] Paul Carrick, op. cit.
[117] The Republic, #407 (Oxford, England: Oxford Univ. Press, 1968 reprint), p. 97.
[118] The Republic, #407, (Oxford, England: Oxford Univ. Press. 1968 reprint), p 99-100.
[119] Averroes, Exposicion de la Republic de Plato ed. M. Cruz Hernandez (Madrid, Spain: Tecnos, 1986), p. 32.
[120] cf. Carroll, James, Constantine's Sword-The Church and the Jews, (New York: Houghton Mifflin Co, 2001). Carroll shows that it was Augustine who stopped the killing of Jews once Christianity became the state religion under Constantine.

Augustine's authority "saved" the Jews from extermination but condemned them to an alienated social status and a continually persecuted existence.

[121] Thomas Aquinas, The Summa Theologica of St. Thomas Aquinas 2.2. Q. 64, a.5, trans. Fathers of the Dominican Province (London: Burns, Oates, and Washbourne, 1929), 202-5.

[122] Nigel Biggar, Aiming to Kill: The ethics of Suicide and Euthanasia, (Pilgrim Press, 2004).

[123] Sir Francis Bacon, quoted by E. Emanuel "The History of Euthanasia Debates in the United States and Britain," Annals of Internal Medicine 121 (15, November, 1994), - 793

[124] Sir Thomas More, Utopia, from The Complete Works of St. Thomas More, ed. E. Surtz and J.H. Hexter (New Haven: Yale University Press, 1963), 4:186.

[125] De Victoria, Francisco, *Relectiones Theologicae* Relectio IX, *De Temperantia*, n. 9 and 12. cf. D. Cronin, The Moral Law in Regard to Ordinary and Extraordinary Means of Conserving Life, (Rome: Gregorian University, 1958), p. 43; Soto, Domingo, De Justicia et Jure (Venice, 1568), Book 5, question 2, article 1. cited by R. Sparks, To Treat or Not to Treat, (New Jersey: Paulist Press, 1988), p. 94-95.

[126] An 1810 French law prohibited killing even patients suffering from rabies, convulsions, and madness.

[127] One might wonder why moral teachings on issues like birth control have been considered by conservative Church officials to be unchangeable.

[128] Sparks, Richard, To Treat or Not to Treat, N.Y., Paulist Press, p. 95.

[129] Task Group of Faith, Worship and Ministry committee, Care in Dying: A consideration of the Practices of Euthanasia and Physician Assisted Suicide, (Continuum Publishing Group, 1999).

[130] Anesthesia in 1848; hypodermic syringe in early 1850s.

[131] cf. Fry, Bruce, "Active Euthanasia: A Historical Survey of its Conceptual Origins and Introduction into Medical Thought," Bulletin of the History of Medicine 52 (Winter, 1978): 492-502.

[132] First Oregon and Washington State passed legislation approving euthanasia and physician assisted suicide. More recently, the state of Montana did likewise. For an analysis of this legislation see Cathleen Kaveny, *Private and Public, Montana's Supreme Court Rules on Assisted Suicide*, (Commonweal, March 12, 2010)

[133] EPEC, Education for Physicians on End of Life Care, is the AMA's effort to change physician behaviors at the end of life.

[134] cf. Bruce Fry p. 449.

[135] K. Binding and A. Hoche, Die Freigabe der Vernichtung lebensunwerten Lebens: Ihr Mass und ihre Form, (Leipzig: F. Meiner, 1920).

[136] A. Jost, Das Recht auf den Tod: Sociale Studie, (Gottingen: Dieterich'sche Verlagsbuchhandlung, 1895).

[137] "The Moral Side of Euthanasia" JAMA, 1885(5): 382-83, quoted by Emanuel, 795.

[138] "It's all over Debbie," JAMA (January 1988): 272

[139] Quill, Timothy, "Death and Dignity: A Case of Individualized Decision Making," NEJOM (March 7, 1991): 691-694.

[140] The AMA sponsored program to train doctors for end of life care is called EPEC: Education for Physicians on End of Life Care.

[141] The U.S. Attorney General recently intervened to restrain or even to eliminate the provision of drugs for suicide by a physician.
[142] Cf. Craig Paterson, Assisted Suicide and Euthanasia: A Natural Law Ethics Approach, (Ashgate Publishing Co, 2008). Contribution of Natural Law Theory to Moral and Legal Debates concerning Suicide, (Booksurge Lic, 2010).
[143] Durkheim in the 19th century defined suicide so broadly that he included martyrdom and heroism under the category. Consequently he concluded that suicide was an acceptable practice in early Christianity. His definition of suicide, however, was so broad as to be confusing and to create a distorted conclusion. Jesus did not commit suicide, and neither did the martyrs who refused to deny their faith; but both would fall under Durkheim's expanded definition: Jesus because he knew that he would be killed and martyrs for the same reason.
[144] In the U.S. things which people want or need are spoken of as rights. In the bioethics literature, people speak of the right to die in this sense. This language is also used in legal decisions. Rights language is used in such broad contexts that the traditional meaning of "right" (a natural or constitutional claim) is lost. If the term right becomes synonymous with a need or desire or interest, then certainly people have a right to suicide. This idea of right was extended in the appellate decisions to become legal rights.
[145] The Supreme Court decision was unanimous. It found that bans on assisting suicide in the Washington and New York Federal Districts do not violate the 14th Amendment. This decision was rendered on June 26, 1997. The decision made clear that states have a right to ban the practice of assisting suicide but also to legalize it. Shortly after, Oregon, Washington state and Montana legalized physician-assisted suicide (2010).
[146] CG Prado, S.J. Taylor, Assisted Suicide: Theory and Practice in Elective Death, (Prometheus Books, 1999).
[147] For an analysis of the issue of limits built into legislation, see D. Orentlichar and L. Snyder, "Can Assisted Suicide be Regulated?" Journal of Clinical Ethics (Winter 2000): 358-366.
[148] The largest demonstration thus far took place on April 10, 2001, the day the Dutch Upper House of Parliament voted 46 to 28 to legalize euthanasia and physician-assisted suicide. The law went into effect when signed by the Queen.
[149] K. Jamison, Night Falls Fast (New York: Alfred A. Knopf, 1999), p. 24.
[150] Ibid. p. 22.
[151] Ibid. 23.
[152] cf. J. Drane, Clinical Bioethics, (Kansas City, MO: Sheed and Ward, 1994), p. 150, 157. James F. Drane, Relief From Suffering and Depression: The Role of Understanding and Faith, (St. Paul's Press, 2010), chapter on suicide.
[153] K. Jamison, Night Falls Fast (New York: Alfred A. Knopf, 1999), p. 24.

Palliative Care For Dying Patients

[154] For a better understanding of this development and the importance of Paracelsus in the 16th Century, see David J. Collins, S.J. Ph.D., God and Nature in Western Thought, (Now You Know Media, 2010).

[155] Von Engelhardt, Dietrich. "Health and Disease: History of the Concepts" in Encyclopedia of Bioethics, Editor, Warren Thomas Reich, Simon and Schuster Mac Millan, N.Y. 1995, p. 1085
[156] Arand, K. J. S. and Carr, D. B. "The Neuroanatomy, Neurophysiology, and Neurochemistry of Pain, Stress, and Analgesic in Newborns and Children" Pediatric Clinics of North America, 36 no.4 p. 795, 822.
[157] Cassell, Eric, " The Nature of Suffering and the Goals of Medicine," NEJOM 306 (1992): 639-45; The Nature of Suffering, N.Y.: Oxford Univ. Press, 1991. Eric Cassell has written extensively about pain and suffering. This paper is endebted to him, both to his conferences on the subject and to articles which appeared before and after his book.
[158] Wall, Patrick. Pain: The Science of Suffering, N.Y., Columbia Univ. Press, 2000, p. 46
[159] For a more humanistic and religious perspective on the meaning of pain and suffering, see Sidney Callahan, Created for Joy – A Christian View of Suffering, (Crossroad Publishing Company, 2007). James F. Drane, Relief From Suffering and Depression: The Role of Understanding and Faith, (St. Paul's Press, 2010).
[160] Byock, Ira R., "End of Life Care: A Public Health Crisis and an Opportunity for Managed Care," The American Journal of Managed Care, v.7 (12) p. 1123, 1132 (December 2001).
[161] Reich, Warren, "Speaking of Suffering: A Moral Account of Compassion," Soundings 72 no.1 p. 83-108. Warren Reich continues to write about the importance of care and compassion. A new book on this topic is anticipated in 2012.
[162] Brand, Paul and Yancy, Philip, Pain, The Gift Nobody Wants, N.Y. Harper Collins Publishers, 1993
[163] Frankl, Victor E., Man's Search for Meaning: An Introduction to Logotherapy. Boston: Beacon Press, 1962; Hawerwas, Stanley, "Reflections or Suffering, Death, and Medicine," Ethics in Science and Medicine, p.229-237 (1979). For more on the issue of meaning in suffering, see James F. Drane, Relief From Suffering and Depression: The Role of Understanding and Faith, (St. Paul's Press, 2010).
[164] The Talmud is a post biblical collection of sayings of rabbis. The sayings are considered to be oral law (as opposed to the written law of the Bible). The oral law recorded in the Talmud is considered a necessary supplement to the written law and of no less value. The teachings and precepts of the Talmud are part of the Jewish sacred scripture. They deal with what today we refer to under the broad category of ethics and law.
[165] Tolstoy, Leon. The Death of Ivan Ilyich, Penguin Books, Baltimore, MD 1960.

TECHNOLOGY

An Ethics of Technology: From the 21st Century

[166] The Supreme Court today is made up of 3 women and 6 men. One male judge is African American. All 9 judges are either Jewish or Catholic. Chief Justice Roberts, Antonin Scalia, Clarence Thomas, Samuel Alito (all Catholics) are the strong conservative members who have generated a rare public criticism from the Obama

Administration for a recent decision removing limits on big business' contributions to politics.

[167] In January 2009, all U.S. media were filled with news about a single woman, already with 6 children, and no job, gave birth to 8 babies. She received in vitro fertilization. The fertility specialist implanted the 8 embryos. Besides the ethical questions raised by the physician's role there were many other questions raised by this reproductive technology. How is this woman with no job and no husband and no income supposed to raise her 14 children? This case shows how new technologies can be misused and how misuse threatens or compromises human life. Cf. Brian Scarnecchia, Bioethics, Law and Human Life Issues: A Catholic Perspective, (Rowman & Littlefield, 2010).

[168] So many of the most basic issues in Bioethics are approached and decided on the basis of our understanding and beliefs about what it means to be human and what is owed morally to human beings. This topic was addressed by the last President's Council on Bioethics. Council members' essays were published by Dr. Edmund Pellegrino who presided over the Council. See Edmund Pellegrino, Human Dignity and Bioethics, (www.bioethics.gov, 2009).

[169] Cf. Thomas A. Shannon, Reproductive Technologies: A Reader, (Rowman & Littlefield, 2004).

[170] Cf. Andrew Lustig, Baruch Brody, Gerald McKinny (editors), Altering Nature: Concepts of Nature and the Natural in Biotechnology Debates, (Springer Verlag, 2008).

[171] Cf. Lan Wilmut, Roger Highfield, After Dolly: The Promise and Perils of Human Cloning, (WW Norton, 2007).

[172] Cf. Roberta Berry, Ethics of Genetic Engineering, (Taylor and Francis, 2007).

[173] Lisa Cahill, Genetics, Theology, and Ethics: An Interdisciplinary Converstion, (Independent Pub Group, 2005).

[174] Aristotle, Nicomachaen Ethics: Book VI, trans. Martin Ostuald (Indianapolis, IL: Bobbs Merrill, Inc. 1962).

[175] Aristotle's Physics is concerned with things which are susceptible to change. In this treatise, he develops the categories for understanding things in the changeable natural world, with special emphasis on causes; cf. Aristotle, Physics, Books I and II.

[176] Summa Theologica, I, II, 94, 3: 21, 1c.

[177] John Duns Scotus, Opus Oxonense, III, 19 and William of Ochham, Quodlibeta, III, 9. 13.

[178] F. Bacon, The New Organon and Related Writings, ed. F.H. Anderson (New York: Classic Books, 1960). The integration of knowledge and power with ethics is seen in one of Bacon's sayings: "Knowledge is power but honesty is authority."

[179] F. Nietzsche, The Will to Power, trans. W. Kaufmann and R.G. Hollingdale, ed. W. Kaufmann, (New York: Random House, 1987).

[180] Historically creativity was understood as a divine characteristic. In the Renaissance geniuses like Michelangelo were thought to be creative. In the 20th century creativity became a characteristic of every young poet, of nursery school finger painters, indeed of everyone who departed from conventional standards. This understanding of creativity is all too easily assigned to scientists, especially the life scientists and

biotechnicians. cf. D. Simonton, Scientific Genius, A Psychology of Science, (Cambridge, England: Cambridge Univ. Press, 1988).

[181] Byran Sykes, Adam's Curse: A Future Without Man, (WW Norton & Sons, 2004).

[182] Bacon refers to this distinction throughout his major work, Novum Organum. Real science, he argues, does more than collect facts. Fact collection alone can end in ignorance. Bacon's method provided a direction for moving beyond fact gathering to real scientific understanding. He rejected Empiricism on one extreme (fact gathering) and Rationalism (classical philosophy) on the other.

[183] Albert Einstein is an example of this broader, more creative, almost mystical involvement with his subject matter.

[184] This point was developed by a great scientist who late in life moved over into the philosophy of science. cf. M.Polanyi, Personal Knowledge, (Chicago, IL: University of Chicago Press, 1962).

[185] The scientists who worked on The Manhattan Project first participated enthusiastically in developing terrible destructiveness but as they progressed they became more and more hesitant. Finally they admitted that they had "sinned." Compared to some politicians and military persons, these scientists showed moral sensitivity.

[186] In May 2000, Dr. James Wilson, a famous geneticist at the University of Pennsylvania killed a young man who was one of his research subjects. It turns out that he had bypassed regulations and ignored limits governing research on human subjects. cf. Washington Post, 25 May 2000, sec. A1, p. 23. A similar tragedy happened at Johns Hopkins University in July 2001. A research subject died after taking an experimental drug. According to FDA investigators, the researcher and the IRB both overlooked dangers related to the experiment.

[187] Harold Baillie, Timothy Casey (editors), Is Human Nature Obsolete: Genetic Engineering and the Future of the Human, (MIT Press, 2004).

[188] cf. L. Walters, Ethics of Human Gene Therapy, (Oxford University Press, 1996). It is worth noting that the two most threatening disease in the world today (AIDS and Mad Cow Disease) originated in the transfer of a microbe from an animal (monkey and cow) to a human being.

[189] cf. B. Joy, "Why The Future Doesn't Need Us," Wired, (April 4, 2000).

[190] The split between what we need from our academic intellectuals and ethicists and what they spend their time doing reminds me of a similar split during the Vietnam period. The whole American society was coming apart because of ethical conflicts over the war, and moral philosophers entrenched in linguistic analysis spent their time parsing words in clauses and phrases.

[191] cf. H. R. Niebuhr, The Responsible Self: An Essay in Christian Moral Philosophy, (New York: Harper & Row, 1963).

[192] Bill Joy, the co-founder and chief scientist of Sun Microsystems, in the above mentioned article in Wired provides examples from genetic engineering, robotics, and nanotechnology of serious threats to human existence and the established order.

[193] Andrew Lustig, Baruch Brody, Gerald McKinny, Altering Nature: Concepts of Nature and the Natural in Biotechnology Debates, (Springer Verlag, 2008).

[194] Alfonso Gomez-Labo, Morality and the Human Good: An Introduction to Natural Law Ethics, (Georgetown University Press, 2002).

[195] It seems ironic that liberals base their moral judgment on an analysis of the structure of the biological reality, following a traditional Catholic Natural Law ethics. The Vatican, the Pope and the bishops on the other hand use a categorical imperative which reflects a Kantian ethics. Maybe some forms of ecumenism are working.

[196] The genetic revolution began with the scientific work of a Catholic monk, Gregor Mendel, in the 19th century (1822-1884). He studied and genetically altered plants. His experiments with plants (peas) mark the beginning of the science of genetics.

[197] I. Berlin, Two Concepts of Liberty, Inaugural Lecture as Chichele Professor of Social and Political Theory , (Oxford, England: 1958) cf. W. McClay "Two Concepts of Secularism," Wilson Quarterly (Summer 2000): 54-71.

[198] M. Minski, "Will Robots Inherit the Earth?," Scientific American (October 1994): 271: 109.

[199] The inherent order of human procreation demands control of the number of offspring that can be cared for and nurtured. Birth control does not violate the inherent order of nature, but having larger families which cannot be adequately cared for does. The octomom and her fertility doctor created a situation which required enormous financial and personal help for her babies to survive. The octomom of 2009 is a good example of ignoring an inherent order of procreation.

Advanced Alzheimer's Disease: Stopping Nutrition and Hydration Technologies

[200] Coincidentally, on February 7, 2001, I saw a program called Family Law which was all about this issue. In American culture, we still have not caught up with the historical Catholic wisdom. When an elderly woman wanted to stop treatment because it was burdensome and provided little benefit and much harm, the case was taken to court and she was denied this choice. Consequently, there was tragedy instead of a humane dying.

[201] Michael Panicola, Making Health Care Decisions: A Catholic Guide to Medically Administered Nutrition and Hydration, (Liquori Pub, 2006).

[202] Christopher Tollefson (editor), Artificial Nutrition and Hydration: The New Catholic Debate, (Springer Verlag, 2007).

[203] T. Finucone, C. Christmas, and K. Travis, "Tube Feeding in Patients with Advanced Dementia," JAMA 282 (Oct. 13, 1999): 1365-1369.

[204] Ronald Hamel, James Walter (editors), Artificial Nutrition and Hydration and the Permanently Unconscious Patient: The Catholic Debate, (Georgetown University Press, 2007).

[205] Stephen Post, Moral Challenge of Alzheimer Disease: Ethical Issues from Diagnosis to Dying, (John Hopkins University Press, 2000).

[206] Pro-Life Committee, National Conference of Catholic Bishops, "Nutrition and Hydration: Moral and Pastoral Reflections," Origins 21 (April 9, 1992): 705-712.

[207] Ibid., p. 710-711.

[208] Texas Conference of Catholic Health Facilities and Texas Conference of Catholic Bishops, "On Withdrawing Artificial Nutrition and Hydration." Origins 21 (Jan. 30, 1992): 541-553.

[209] Pennsylvania Conference of Catholic Bishops. "Nutrition and Hydration: Moral Considerations." Origins 21 (Jan. 30, 1992).
[210] Pro-Life Committee, p. 711.
[211] Ibid., p. 428
[212] Ibid., p. 707
[213] "Declaration of Helsinki: Recommendations Guiding Medical Doctors in Biomedical Research Involving Human Subjects," 1964, revised 1975. In R. Levine, Ethics and Regulation of Clinical Research, Second Edition (Baltimore, MD: Urban and Schwarzenberg, 1986,) p. 427-429.
[214] Ibid., 428.
[215] Pro-Life Committee, p. 710.
[216] Ibid., p. 711
[217] L. Schneiderman, N. Jecker, and A. Jonsen, "Medical Futility: Its Meaning and Ethical Implications," Annals of Internal Medicine 112 (June 15, 1990): 949-954.
[218] J. Drane, and J. Coulehan, "The Concept of Futility," *Health Progress*, (December, 1993). Charlelle Gallager-Alfred, Madalon Amenta (editors), Nutrition and Hydration in Hospice Care: Needs, Strategies, Ethics, (Taylor and Francis, 1997).

Equity and Justice in High Tech Medicine

[219] James F. Drane, Pilgrimage to Utopia, (Bruce Publishing Co, 1965).
[220] It seems ironic, but today in China, the government system is too expensive and officials have opted for a private free-market system to ease the economic pressure. The value sacrificed in this change of course is equality.
[221] Edmund Pellegrino, Philosophy of Medicine Reborn: A Pellegrino Reader, (University of Notre Dame Press, 2008).
[222] R. Veatch, "Justice, The Basic Social Contract and Health Care", in *Contemporary Issues in Bioethics*, T. Beauchamp, L. Walkers (eds.), (Belmont, California: Wadsworth Publishing Co. 1989).
[223] Norman Daniels, James Sabin, Setting Limits Fairly: Learning to Share Resources for Health, (Oxford University Press, 2008).
[224] This is one of the central points in Daniel Callahan's latest book, Taming the Beloved Beast: How Medical Technology Costs Are Destroying Our Health Care System, (Princeton University Press, 2009).
[225] Norman Daniels, Bruce Kennedy, Ichiro Kawachi, Is Inequality Bad For Our Health, (Lightning Source, 2000).
[226] Therese Murphy (editor), New Technologies and Human Rights, (Oxford University Press, 2009).
[227] Richard Ashcroft, Angus Dawson, Heather Draper (editors), Principles of Health Care Ethics, (John Wiley & Sons Inc, 2007).
[228] J. Rawls, A Theory of Justice, (Cambridge, Mass.: Harvard University Press, 1971).
[229] President Obama's Health Reform bill pushed first toward a broad equal access provided by a Federal Government plan similar to Medicare. Republican opposition removed this part of his plan. Following the Obama Health Care bill through the

House and Senate one sees all the challenges that are raised to equity in health care (2010).

[230] C. Sykes, A Nation of Victims; The Decay of the American Character, (New York: St. Martins Press, 1982).

[231] N. Daniels, Just Health Care, (Cambridge, MA: University Press, 1985).

[232] J. Ortega y Gasset, Obras Competa, (Madrid: Alianza, 1946).

[233] J. Drane, Becoming a Good Doctor, The Place of Virtue and Character in Medical Ethics, (Kansas City, MO: Sheed & Ward, 1995). Relief From Suffering and Depression: The Role of Understanding and Faith, (St. Paul's Press, 2010).

Ethik in der Praxis/Practical Ethics
Studien/Studies
hrsg. von Prof. Dr. Hans-Martin Sass (Universität Bochum/Georgetown University Washington)
Schriftleitung: Dr. Arnd T. May

Georg Bollig
Palliative Care für alte und demente Menschen lernen und lehren
Viele alte und demente Menschen brauchen Palliative Care. Das Ziel des hier beschriebenen Allgemeinwissen-Ansatzes und der Kette der Palliativversorgung ist die Verbreitung von Wissen um Palliative Care in der gesamten Bevölkerung und die Verfügbarkeit für ALLE, die es brauchen. Innovative Ausbildungsmodelle und ein Stufenmodell für Lernen und Lehren werden vorgestellt. Inhalte eines Basiskurses in Palliative Care für alle Mitarbeiter in Altenhilfe und Gesundheitswesen werden aufgezeigt. Kurse in „Letzte Hilfe“ könnten ein Basiswissen in Palliative Care in der gesamten Bevölkerung verbreiten.
Bd. 35, 2010, 128 S., 19,90 €, br., ISBN 978-3-643-90058-6

LIT Verlag Berlin – Münster – Wien – Zürich – London
Auslieferung Deutschland / Österreich / Schweiz: siehe Impressumsseite

Johannes Huber
Totipotenz – überfordertes Kriterium der Schutzwürdigkeit?
Eine naturphilosophische Untersuchung zu den biologischen Grundlagen eines normativ gewordenen Begriffs
Geleitwort Prof. em. Dr. med. Dr. rer. nat. Henning M. Beier
LIT